Workbook 2

Delivering the AQA Specification

NEW GCSE MATHS
AQA Modular

Matches the 2010 GCSE Specification

• Chris Pearce •

CONTENTS

RECALL

UNIT 3

INTRODUCTION

Welcome to Collins New GCSE Maths for AQA Modular Workbook 2. The first part of this book covers Recall content which you will have learnt in Unit 1 and Unit 2. You will also need some content covered in this section for your Unit 3 exam. The second part covers the content that is specific to Unit 3.

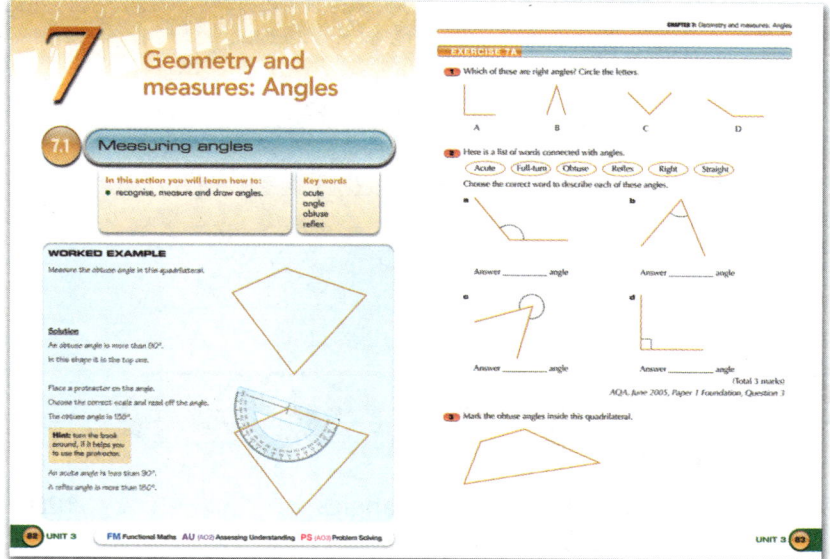

Worked examples

Understand the topic by reading the examples in blue boxes. These take you through questions step by step.

Colour-coded questions

Make progress as you move from red to orange to yellow questions.

Exam practice

Prepare for your exams with past exam questions.

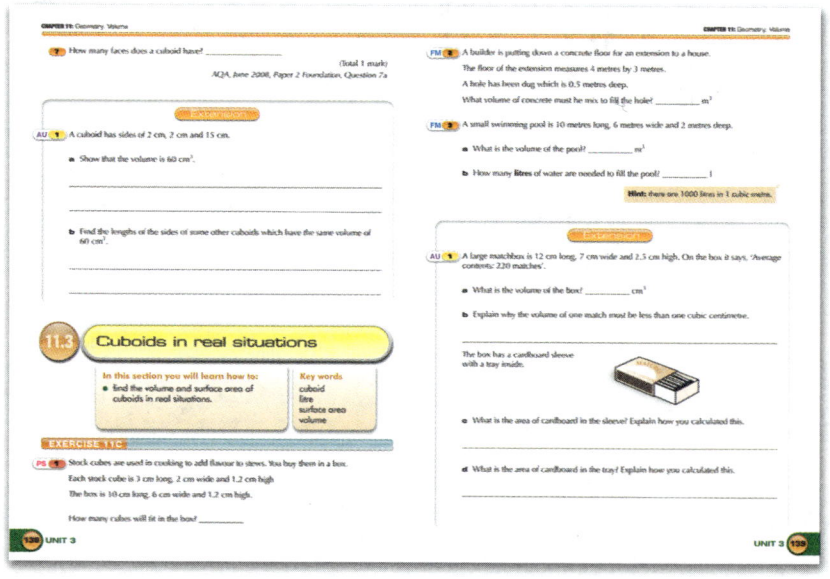

Apply your maths

Practise maths questions in a range of situations in the separate section in yellow at the end of each chapter.

New Assessment Objectives

Check how well you have understood each topic with questions that assess your understanding marked **AU** and questions that test if you can solve problems marked

 .

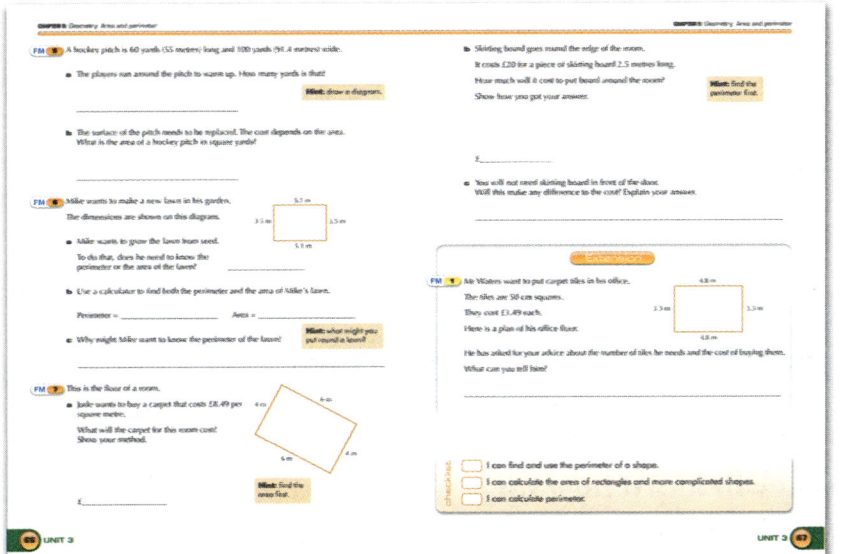

Checklist

Tick off the topics you can do on the checklist at the end of each chapter.

Helpful hints

Use hint boxes to give you tips as you work through the questions.

Functional maths

Practise functional maths skills to see how people use maths in everyday life. Look out for practice questions marked **FM** and there are also extra functional maths and problem-solving activities at the end of every chapter.

Answers

Check your own work with the answers at the back of the book.

Glossary

Look up unfamiliar words in the glossary at the back.

1 Number: Using a calculator

1.1 Arithmetic with a calculator

In this section you will learn how to:
- use a calculator efficiently.

Key words
average
calculator
fraction

WORKED EXAMPLE

The ages of five people are 23, 28, 37, 41 and 46.

What is their average age?

<u>Solution</u>

To find the average, we add up the five ages and divide by 5.

We can do that with a calculator, using brackets like this:

$(23 + 28 + 37 + 41 + 44) \div 5 =$

On most calculators you also do it by pressing the keys shown here:

23 + 28 + 37 + 41 + 44 [=] [÷] 5 [=]

Check you can do it both ways on your calculator to get the answer 36.

EXERCISE 1A

Use a calculator to answer these questions.

 **1** Four elderly ladies said that their total age was 299 years.

Three of them were 82, 67 and 66.

How old was the fourth one?

_____

FM Functional Maths   **AU** (AO2) Assessing Understanding   **PS** (AO3) Problem Solving

2 Find half the sum of 654 and 978.

_____

3 Here are the temperatures on seven days in one week.

   14   17   16   19   18   11   10

What is the average for the week?   **Hint:** you must divide by 7.

_____

4 Ann has £185, Jean has £247 and Marie has £219.

They decide to put all their money together and share it out equally.

How much will each person get?

_____

FM 5 Yasmin is saving £15 a week from her salary.

How long will it take her to save £250?   _____

Explain how you know.

_____

AU 6 Work out the following:

$26 \times 26 - 27 \times 25 =$ _____

$43 \times 43 - 44 \times 42 =$ _____

$87 \times 87 - 88 \times 86 =$ _____

What do you notice?

_____

_____

Write another multiplication like the ones above.

_____

7 There are 60 minutes in an hour, 24 hours in a day and 365 days in a year.

   a How many minutes are there in a day?   _____

   b How many hours are there in a year?   _____

Extension

PS 1 Two of the digits are missing in this multiplication.

Use a calculator to help you find them.

$$\boxed{}6 \times \boxed{}4 = 1944$$

PS 2 Find three different odd numbers, all larger than 5, which multiply to make 1001.

$$\boxed{} \times \boxed{} \times \boxed{} = 1001$$

AU 3 Carlos and Juan have 476 euros between them.

Carlos has 50 euros more than Juan.

How much does each one have?

_____

1.2 Fractions with a calculator

In this section you will learn how to:
- use a calculator to work with fractions.

Key words
fraction
mixed number

WORKED EXAMPLE

What is $\frac{3}{5} + \frac{7}{10}$?

Solution

Use the fraction button on your calculator.

It may give the answer as $\frac{13}{10}$ or as $1\frac{3}{10}$.

Make sure you know how to use your calculator with fractions.

EXERCISE 1B

1 Use the fraction key on your calculator to simplify these fractions.

a $\dfrac{8}{12} =$ _____   **b** $\dfrac{18}{24} =$ _____

c $\dfrac{18}{30} =$ _____   **d** $\dfrac{35}{40} =$ _____

2 Draw lines to join the equivalent fractions.

$$\frac{3}{6}$$ $$\frac{4}{6}$$ $$\frac{7}{21}$$

$$\frac{12}{18}$$ $$\frac{9}{27}$$ $$\frac{5}{10}$$

3 Find four fractions which are equivalent to $\dfrac{28}{40}$.

_____

_____

4 Add these fractions with a calculator.

a $\dfrac{1}{2} + \dfrac{1}{3} =$ _____   **b** $\dfrac{1}{4} + \dfrac{3}{8} =$ _____   **c** $\dfrac{1}{3} + \dfrac{1}{4} =$ _____

d $\dfrac{3}{4} + \dfrac{1}{8} =$ _____   **e** $\dfrac{1}{2} + \dfrac{1}{6} =$ _____   **f** $\dfrac{2}{3} + \dfrac{1}{6} =$ _____

PS 5 $\dfrac{1}{4} + \dfrac{1}{4} = \dfrac{1}{2}$

Can you find two **different** fractions which add up to $\dfrac{1}{2}$?

_____

The fraction $\frac{7}{4}$ can be written as $1\frac{3}{4}$. The second way is called a **mixed number**. Make sure you can show this using a calculator.

6 Use a calculator to write these fractions as mixed numbers.

a $\frac{5}{4} =$ _____

b $\frac{7}{3} =$ _____

c $\frac{11}{4} =$ _____

d $\frac{23}{4} =$ _____

e $\frac{13}{3} =$ _____

f $\frac{17}{6} =$ _____

7 Add these fractions and give your answer as a mixed number. Use a calculator.

a $\frac{3}{4} + \frac{1}{2} =$ _____

b $\frac{1}{2} + \frac{2}{3} =$ _____

c $\frac{3}{4} + \frac{5}{8} =$ _____

d $\frac{7}{8} + \frac{1}{2} =$ _____

e $\frac{5}{8} + \frac{5}{8} =$ _____

f $\frac{7}{8} + \frac{7}{8} =$ _____

AU 8 **a** Explain how this clock face shows that $\frac{1}{4} + \frac{1}{3} + \frac{5}{12} = 1$.

_____

_____

b Check the addition is correct using your calculator.

9 Use your calculator to do these subtractions.

a $1 - \frac{1}{3} =$ _____

b $\frac{7}{8} - \frac{1}{4} =$ _____

c $1 - \frac{3}{4} =$ _____

d $\frac{5}{8} - \frac{1}{4} =$ _____

e $2 - \frac{3}{4} =$ _____

f $2 - \frac{2}{3} =$ _____

10 Fill in this addition table for fractions. Use a calculator to help you.

+	$\frac{1}{4}$	$\frac{1}{3}$	$\frac{1}{2}$	$\frac{3}{4}$
$\frac{1}{4}$				
$\frac{1}{3}$				
$\frac{1}{2}$				
$\frac{3}{4}$				$1\frac{1}{2}$

Extension

AU **1** Calculate $\frac{3}{8} - \frac{1}{3} =$ _____.

How does this show that $\frac{3}{8}$ is bigger than $\frac{1}{3}$?

_____

AU **2** Which is bigger, $\frac{2}{3}$ of a cake or $\frac{5}{8}$ of a cake? Explain how you know.

_____

_____

3 Use your calculator to find the following.

a $1\frac{1}{2} + 2\frac{3}{4} =$ _____   b $3\frac{1}{2} - 1\frac{3}{4} =$ _____   c $\frac{1}{2}$ of $5\frac{1}{2} =$ _____

1.3 Calculator problems

In this section you will learn how to:
● use a calculator to solve practical problems.

Key words
average
fraction

EXERCISE 1C

1 This was the population of Margate in Kent at the time of the 2001 census.

Males	27 630
Females	30 835

a What was the total population of Margate? _____

b There were more females than males. How many more? _____

2 Complete this shopping bill.

4 oranges at 26p each	
$\frac{1}{2}$ kg of grapes at £3.20 per kg	
Total	

(Total 3 marks)

AQA, June 2008, Paper 2 Foundation, Question 1

FM 3 Elise has four test marks on her college course.

They are 72, 49, 58 and 83.

To pass the course, she must have an average of at least 65. To find the average, she must add the marks and divide the answer by 4.

Has she passed? Give a reason for your answer.

_____

_____

PS 4 Here is a table of costs for hiring a car. The cost is in pounds per day.

Vehicle type	Up to 7 days	8 to 14 days	15 or more days
Compact	46	44	42
Intermediate	57	53	48
Station wagon	68	62	57

a Find the cost of a compact car for 13 days. _____

b Find the cost of a station wagon for 24 days. _____

c What is the difference in cost between a
station wagon and a compact car for a 9 day hire? _____

PS **5** Here are the costs of a range of kitchen units.

Unit	Cost in pounds
Tall wall unit	56
Short wall unit	72
Cupboard unit	128
Drawer unit	156

Find the total cost of two tall wall units, two short wall units, three cupboard units and a drawer unit.

_____

FM **6** Andy and Becca want to buy some furniture. Here are some prices.

Furniture	Cost
Armchair	£795
Small sofa	£995
Medium sofa	£1145
Large sofa	£1275

They are thinking about two options.

A   Two armchairs and a medium sofa.

B   A large sofa and a small sofa.

Which option is cheaper and by how much?

_____

7 The cost of sending airmail letters is shown.

Weight not over	Europe cost (£)	Asia cost (3)
10 g	0.44	0.50
20 g	0.44	0.72
40 g	0.64	1.12
60 g	0.83	1.51
80 g	1.02	1.91
100 g	1.21	2.31

a Find the cost of sending one letter weighing 45 g to Europe.

Answer £_____ (1)

b Find the total cost of sending two letters, each weighing 82 g, one to Europe and one to Asia.

Answer £_____ (2)

c Find the total cost of sending three letters, each weighing 26 g, to Europe.

Answer £_____ (2)

(Total 5 marks)

AQA, June 2008, Paper 2 Foundation, Question 6

8 A water meter at a house records the volume of water used, in cubic metres. The metre readings at the start and end of a 3 month period are as follows.

	Reading in cubic metres
End	4205
Start	4154

Water costs 104p per cubic metre.

Find the cost of the water used in this period.
Give your answer in pounds.

Answer £_____

(Total 4 marks)

AQA, June 2008, Paper 2 Foundation, Question 17a

Extension

1 You can use a calculator to find fractions of a quantity.

For example, to find $\frac{2}{3}$ of £84 you calculate:

$$\frac{2}{3} \times 84 = 54$$

The answer is £54. This means that in a sale like the one advertised here, the sale price is $\frac{2}{3}$ of the original price.

Complete this table. The first item is done for you.

Item	Original price	Sale price
Camera	£84	£54
Headphones	£57	
Game console	£177	
Television	£468	

FM 2 The price of a computer is £396. The price in a sale is $\frac{1}{4}$ off if you buy it online, but there is a £45 delivery charge.

How much will it cost altogether to buy online?

_____

checklist

☐ I can use a calculator efficiently.

☐ I can use a calculator to work with fractions.

☐ I can use a calculator to solve practical problems.

Problem Solving
Making jewellery

Jewellery can be made from lengths of wire and beads.

Beads are sold in different sizes.

Wire is sold in different thicknesses, called the gauge.

To make a piece of jewellery the beads are threaded onto the wire.

Getting started

Step 1 Choose a gauge of wire and cut the length required.

Step 2 Put a fastener on one end.

Step 3 Thread beads of different sizes into a pattern.

Step 4 Put a fastener on the other end. Your jewellery is now complete.

Hint: for bracelets and anklets use 20-gauge wire. For necklaces use 24-gauge wire.

Beads are available in three lengths:
● 6 mm ● 8 mm ● 10 mm

Beads are available in four colours:
● green ● blue ● red ● yellow

Task A

1 How many 6 mm beads are needed to make a bracelet?

2 How many 8 mm beads are needed to make an anklet?

3 How many 10 mm beads are needed to make a short necklace?

4 How many 10 mm beads are needed to make a long necklace?

Task B

You are asked to make a bracelet with beads of two different lengths.

You decide to use 6 mm red beads and 8 mm blue beads.

How many would you need if you used them alternately?

Standard lengths for bracelets and necklaces	
Bracelet	17 cm
Anklet	23 cm
Short necklace	39 cm
Long necklace	46 cm
Remember 1 cm = 10 mm	

Task C

Design a piece of jewellery of your choice.

Make a list of all the materials you need. Specify the number of beads and length of wire.

Task D

Work out the cost of your design.

Use the information in the table below.

Price	
6 mm beads	10p each
8 mm beads	12p each
10 mm beads	15p each
24-gauge wire	10p per cm
20-gauge wire	8p per cm
Fasteners for both ends	30p per item of jewellery

2 Measures: Units

2.1 Units of measurement

In this section you will learn how to:

- convert metric and non-metric units.

Key words

approximately
centimetre
gram
kilogram
kilometre
litre
metre
mile
millilitre
millimetre
pint
pound

You need to **remember** the conversions between these **metric units**.

10 mm = 1 cm	100 cm = 1 metre (m)	1000 m = 1 kilometre (km)
1000 mg = 1 gram (g)	1000 g = 1 kg	1000 kg = 1 tonne
1000 ml = 1 litre (l)		

Here are some useful conversions between **metric** units and **non-metric** units.

8 kilometres is approximately 5 miles
1 mile is approximately 1.6 kilometres
1 stone is approximately 6.3 kilograms

1 litre is approximately $1\frac{3}{4}$ pints
4 litres is approximately 7 pints
1 kilogram is approximately 2.2 pounds

WORKED EXAMPLE

In France, a road sign shows that the distance to Paris is 160 km. How many miles is that?

Solution

It is useful to remember that 8 km is approximately 5 miles.

So 80 km is approximately 50 miles   → Multiplying by 10.

and 160 km is approximately 100 miles   → Doubling the numbers.

You could also remember that 1 km is just over half a mile.

So 160 km will be a bit more than 80 miles.   → 80 is half of 160.

This is less accurate but gives you a reasonable approximation to the answer.

EXERCISE 2A

1 Write down a sensible **metric** unit to measure the following.

   a The length of your foot _____   **b** The capacity of a glass _____

   c The weight of a baby _____   **d** The distance between two towns _____

   e The weight of a spoonful of sugar _____   **f** The length of a room _____

2 Fill in the missing numbers.

   a 90 mm = _____ cm   **b** 4 m = _____ cm

   c 2 km = _____ m   **d** half a metre = _____ cm

   e half a kilometre = _____ m   **f** 5.3 cm = _____ mm

3 Fill in the missing numbers.

   a 2 litres = _____ ml   **b** half a litre = _____ ml   **c** $\frac{1}{4}$ of a litre = _____ ml

4 Which metric unit would you use to measure the following?

   a The length of a pencil _____ (1)   **b** The amount of petrol in a car's tank _____ (1)

   c The area of a football pitch _____ (1)   **d** The weight of a bus _____ (1)

   (Total 4 marks)

   AQA, June 2007, Paper 1 Foundation, Question 6

5 Write these lengths in metres.

   a 3.5 km = _____ m   **b** 4.65 km = _____ m   **c** 0.35 km = _____ m

6 Write these quantities in ml.

   a 1.5 litres = _____ ml   **b** 2.85 litres = _____ ml

   c 0.61 litres = _____ ml

Some non-metric units are still in common use.
Miles, pints and stone (for weight) are three examples.

Hint: use the approximate conversions at the start of this section to calculate these.

7 Change these lengths to kilometres.

a 10 miles = _____ km

b 20 miles = _____ km

c 3 miles = _____ km

8 Change these quantities to pints.

a 2 litres = _____

b 8 litres = _____

c Is half a litre more or less than a pint? _____

9 People often know their weight in stones.

Change these weights to kilograms.

a 2 stones = _____

b 5 stones = _____

c 10 stones = _____

Extension

PS 1 A pint is exactly 576 ml.

a How many millilitres are in 2 pints? _____

b How many litres is that? _____

c A gallon is 8 pints. How many litres is 1 gallon? _____

AU 2 Work with a friend on this question.

The weight of a small car is about 1 tonne. Elephants weigh several tonnes.
A group of teenagers find that their total weight is exactly 1 tonne.

Estimate how many teenagers are in the group. Justify your answer.

_____

_____

2.2

Time and timetables

In this section you will learn how to:
- carry out calculations involving time and timetables.

Key words
24-hour clock

WORKED EXAMPLE

A train journey starts at 1735 and finishes at 2112. How long does it last?

Solution

The best way is counting on.

Method 1

1735 to 1800 is 25 minutes.

1800 to 2100 is 3 hours.

2100 to 2112 is 12 minutes.

The total time is 25 minutes + 3 hours + 12 minutes = 3 hours and 37 minutes.

Method 2

1735 to 2035 is 3 hours.

2035 to 2100 is 25 minutes.

2100 to 2112 is 12 minutes.

The total time is 3 hours + 25 minutes + 12 minutes = 3 hours and 37 minutes.

Be careful if you use a calculator. 2112 − 1735 will <u>not</u> give the correct answer because there are 60 minutes in an hour, not 100.

EXERCISE 2B

1. Bus and train timetables are written using the 24-hour clock.

   9.30 am is written as 0930.

   9.30 pm is written as 2130.

   Change these to 24-hour clock times.

   a 8.15 am is _____
   b 1.20 pm is _____

   c 5.00 pm is _____

2 If the clocks below show times in the morning, write them as 24-hour clock times.

a _____ b _____ c _____

3 If the clocks in question 2 show times in the afternoon or evening, write them as 24-hour clock times.

a _____ b _____ c _____

4 How long are these time intervals?

a 1500 to 1534 _____ b 0900 to 0909 _____

c 1340 to 1400 _____ d 1633 to 1700 _____

e 1805 to 1900 _____ f 0713 to 0800 _____

5 Here is part of a bus timetable.

a How frequently do buses leave Adderley?

Adderley	1324	1354	1424	1454	1524
Binford	1345	1415	1445	1515	1545
Chatterton	1355	1425	1455	1525	1525
Denbigh	1411	1441	1511	1541	1541

_____

How long do these journeys take?

b Adderley to Binford _____ c Binford to Chatterton _____

d Chatterton to Denbigh _____ e Adderley to Denbigh _____

Martin arrives at the Adderley bus stop at 2 pm.

f How long will he wait for a bus? _____

g He wants to get to Denbigh by 2.30 pm. How late will he be? _____

6 How long are these time intervals?

a 1015 to 1035 _____

b 1255 to 1310 _____

c 1930 to 2010 _____

d 0720 to 0820 _____

e 1610 to 1735 _____

f 1140 to 1315 _____

Extension

1 **a** A long train journey starts at 0943 and finishes at 1422.

How long did it take? _____

b Another train journey starts at 1325 and lasts 4 hours and 40 minutes.

At what time did it end? _____

c A third journey lasted two hours and thirty minutes and finished at 1115.

At what time did it start? _____

d Write your answers to **b** and **c** as 12-hour clock times using am or pm.

_____   _____

2.3 Units in action

In this section you will learn how to:
● use different units in real settings.

Key words
approximately

EXERCISE 2C

You may wish to use a calculator in this section.

1 Three of the races in the athletics in the Olympics are 1500 metres, 5000 metres and 10 000 metres. Write these lengths in kilometres.

PS **2** You can buy milk in cartons in metric or non-metric measurements.

Put these four cartons in order of size, smallest first:

1 pint, 0.5 litres, 2 pints, 1 litre

_____

3 A magazine has an article about a woman who has been on a diet.
It says that she has reduced her weight from 14 stone to 10 stone.

a Approximately, what are those two weights in kilograms? _____

b How much weight did she lose (approximately) in kilograms? _____

4 Large juice cartons usually hold 1 litre.

Small individual juice cartons often hold 200 ml or 250 ml.

How many of these smaller cartons are equivalent to one large one?

_____

5 Sugar used to be sold in 2 pound bags. Now it is sold in 1 kg bags.

If 1 pound = 454 grams, what is the difference between those two weights?

_____

AU **6** Petrol used to be sold in gallons. Now it is sold in litres.

A litre is approximately 0.22 gallons.

A medium-sized car fuel tank has a capacity of about 50 litres. How many gallons is that?

_____

7 Here are the times for a train from
Birmingham to Glasgow.

a How long is the whole journey?

_____

Birmingham	0920
Crewe	1009
Preston	1051
Carlisle	1200
Glasgow	1319

b The stops split the journey into four parts. How long will each part take?

_____

c Which pair of consecutive stations are likely to be the furthest apart?

_____

Extension

AU **1** A marathon is just over 26 miles long.

Roughly how many kilometres is that? _____

Hint: remember that 1 mile is approximately 1.6 km.

FM **2** Builders always put all lengths on plans and drawings in millimetres.

Change these dimensions into metres.

a The length of a room is 5400 mm _____

b The height of a ceiling is 2850 mm _____

c The length of a window is 1800 mm _____

d The height of a kitchen worktop is 930 mm _____

e Why do you think builders always write lengths in mm?

_____

_____

3 Here is a train timetable

Cleethorpes	0528	0628	0714	0801
Scunthorpe	0600	0700	0745	0832
Doncaster (arr)	0632	0733	0818	0908
Doncaster (dep)	0636	0738	0820	0915
Meadowhall	0701	0805		0941
Sheffield	0710	0814	0852	0950

a How long does the 0528 train from Cleethorpes take to travel to Scunthorpe?

_____ minutes   (1)

b Anna is travelling from Doncaster to Sheffield.
She needs to be in Sheffield by half past nine in the morning
Which train is the latest from Doncaster she should catch?

_____   (1)

c The 0714 train from Cleethorpes to Sheffield has the shortest time of these four trains.
Give a possible reason for this.

_____

_____   (1)

(Total 3 marks)

AQA, March 2009, Module 3 Foundation, Question 13

checklist

☐ I can convert metric and non-metric units.

☐ I can carry out calculations involving time and timetables.

☐ I can use different units in real settings.

You could work with a partner on these questions.

Metric units only came into regular use in England in the 20th century.

This section looks at some of the units that used to be used to measure length.

This railway bridge at Keynsham is 112 **miles** and 63 **chains** from Paddington station. Miles and chains were used to measure distances when the railways were built in the 19th Century.

Getting started

Did you know:
12 inches = 1 foot
3 feet = 1 yard
1760 yards = 1 mile

Have you heard of **chains** and **furlongs**?

1 furlong = 220 yards

The lengths of horse races are still usually given in furlongs. The posts along the track are called furlong posts, because they mark out this distance.

A furlong (a 'furrow length') was the distance a team of oxen could plough without resting.

1 chain = 22 yards

The distance between the wickets in a cricket match is one chain.

The chain takes its name from a real chain that was first used to survey land accurately in 1620 by Edmund Gunter.

Distances in chains are still found in many real estate records.

A chain is approximately 20 metres.

Task A

1 How many inches make one yard?
2 How many feet make one mile?
3 How many chains make one furlong?
4 How many furlongs make one mile?
5 What fraction of a mile is one furlong?
6 How many chains make one mile?

Task B

1 How many metres is a furlong?
2 How many furlongs make one kilometre?
3 Think of some distances around your school that could sensibly be measured in chains. Make a list of estimated distances.
4 Think of some distances in your neighbour-hood that could sensibly be measured in furlongs. Make a list of estimated distances.

A **quarter of a chain** has several names: a **rod**, a **pole**, or a **perch**. It was originally the length of the stick that a ploughboy used to control the oxen pulling the plough.

5 How many feet make one rod?
6 How many rods make one chain?

Task C

You have been given information about a number of old units of measurement including furlongs, chains and rods.

You have also been given the metric equivalents of some of them.

Construct a table that could be used to look up the connections between different old units and to find their approximate metric equivalents.

You might like to make this into a poster and include illustrations.

3 Geometry and measures: Scale and drawing

3.1 Scale

In this section you will learn how to:
- read scales and estimate distances.

Key words
estimate
scale

WORKED EXAMPLE

This scale is marked in grams.

What is the weight shown on the scale?

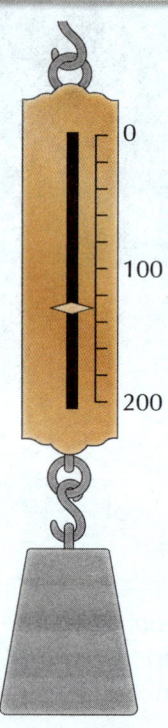

Solution

Look carefully at the scale.

There are five intervals between 0 and 100, ————————————→ $5 \times 20 = 100$
so it must be going up in 20s.

The pointer is halfway through the second interval after 100.

The weight is $100 + 20 + 10 = 130$ g.

FM Functional Maths **AU** (AO2) Assessing Understanding **PS** (AO3) Problem Solving

EXERCISE 3A

1 What temperatures are shown on these medical thermometers?

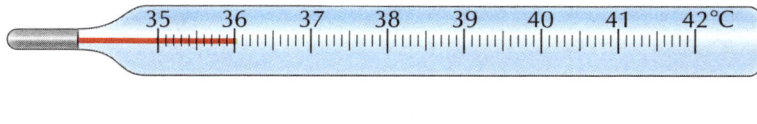

a _____

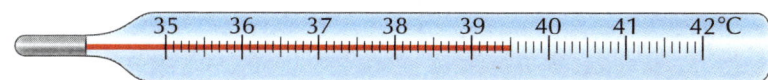

b _____

c _____

2 A normal body temperature is 37.0 °C.
A fever is a temperature of 37.7 °C or above.

Show these temperatures on the thermometers.

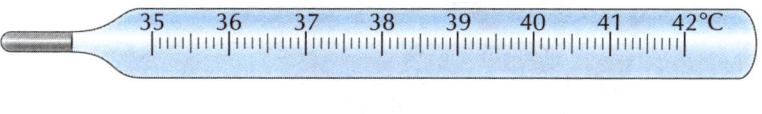

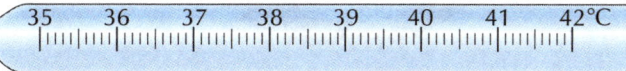

3 What speeds are shown on these car speedometers?

a _____   b _____   c _____

4 Give the values shown by the arrows on these scales.

a

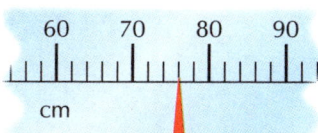

Answer _____ cm   (1)

b

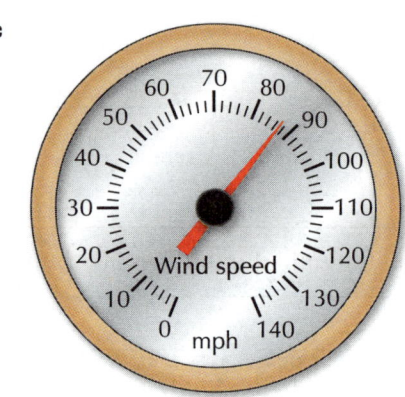

Answer _____ kg   (1)

c

Answer _____ mph   (1)

(Total 3 marks)

AQA, June 2005, Paper 2 Foundation, Question 8

5   What quantity of liquid is shown in these jugs? Give your answers in ml.

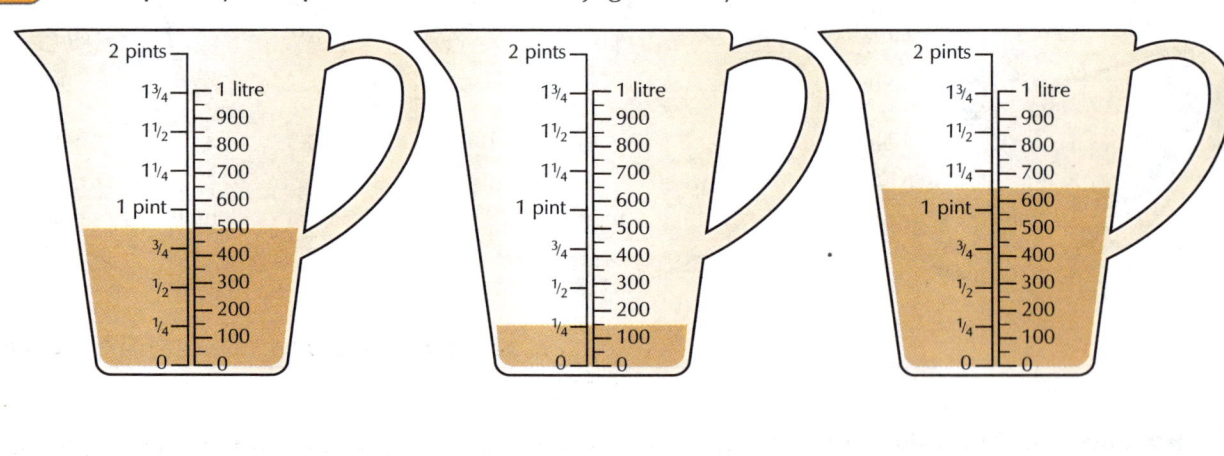

a _____   **b** _____   **c** _____

6 **a** Write down the value shown by the arrow on each of these scales.

i

Answer _____ kmph   (1)

ii

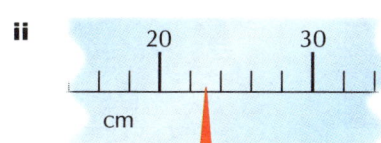

Answer _____ cm   (1)

b The scale below measures mass in grams.
Draw an arrow against the point on the scale that shows 560 grams.

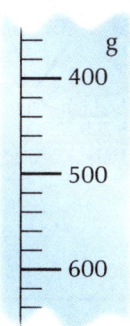

(1)

(Total 3 marks)

AQA, June 2007, Module 5, Paper 2 Foundation, Question 1

7 The car in this picture is 4 m long. Estimate the lengths of the bicycle and the bus.

Bicycle: _____ m   Bus: _____ m

8 The standard height of a door is 2 m.
Estimate the height of the ceiling in the room you are in.   _____ m

Extension

PS **1** Find the weight of each of the three parcels on the scales below.

Show how you found them.

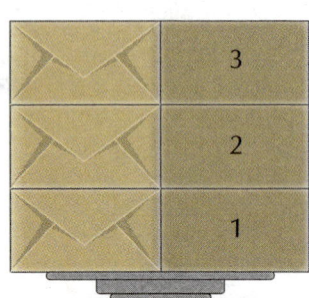

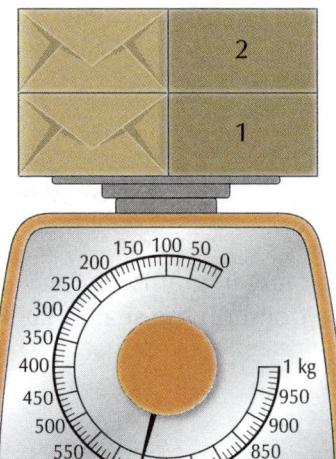

_____

_____

AU **2** An old recipe has liquid quantities in pints and you wish to change them to millilitres.

Use the scale on the jug to complete the table below.

Pints	$\frac{1}{4}$	$\frac{1}{2}$	$\frac{3}{4}$	1	$1\frac{1}{4}$
millilitres					

3.2 Nets

In this section you will learn how to:
● recognise nets of different 3D objects.

Key words

cube	pyramid
cuboid	prism
cylinder	net

WORKED EXAMPLE

Complete this net for a cuboid.

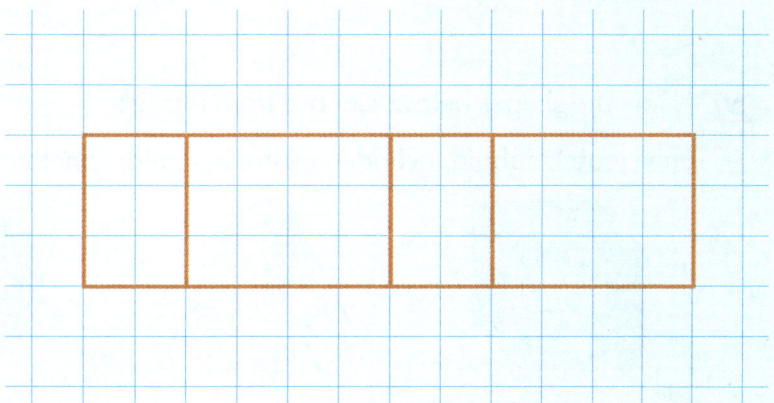

Solution

A cuboid is a rectangular box.

Four faces are shown and two are missing.

The four faces can fold up like this. ———————➤ If you are unsure, cut out a copy on squared paper.

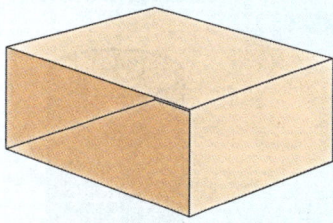

The missing faces are on the front and back.

You could add them like this: ———————➤ There are other ways to do this.

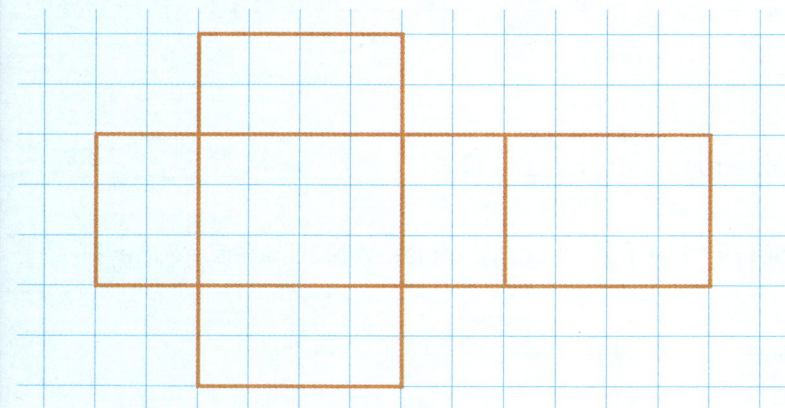

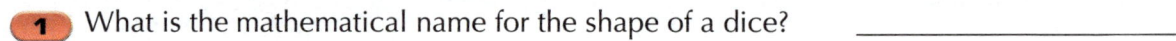

EXERCISE 3B

1 What is the mathematical name for the shape of a dice? _____

2 If you put three dice together like this, what shape do they make?

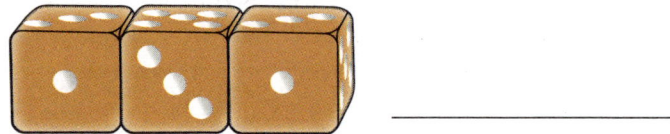   _____

3 Name the shapes below. Choose from this list.

cone, cube, cuboid, cylinder, prism, pyramid, sphere

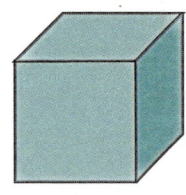         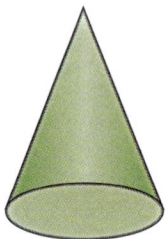

a _____   **b** _____   **c** _____   **d** _____

      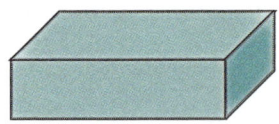

e _____   **f** _____   **g** _____

AU 4 a What is the mathematical name for a baked bean can? _____

b A baked bean can is made out of three flat pieces of metal. What shapes are they?

_____

5 Here is a net.

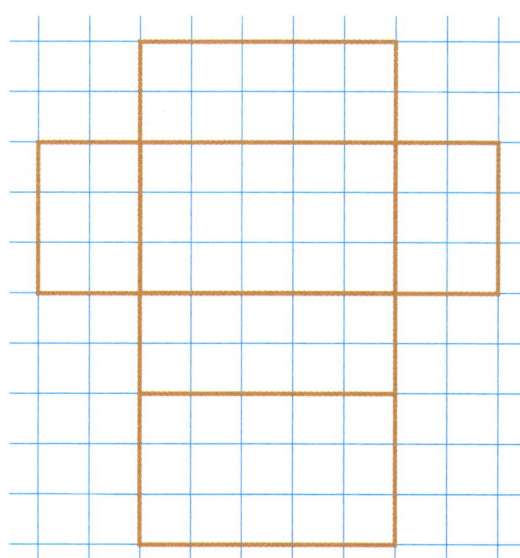

a Circle the name of the shape it makes from this list:

      cube   pyramid   cuboid   prism   cylinder

b Draw here a **different** net that would make the **same** shape.

6 Here is an incomplete net for a pyramid.

One face is missing. Draw it in.

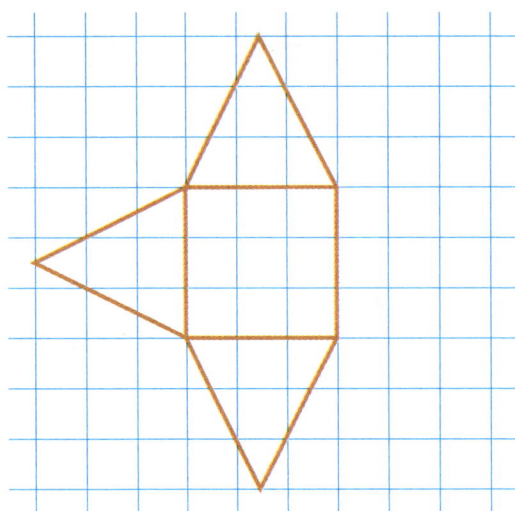

7 Four of these are nets for a cube. Which ones? _____

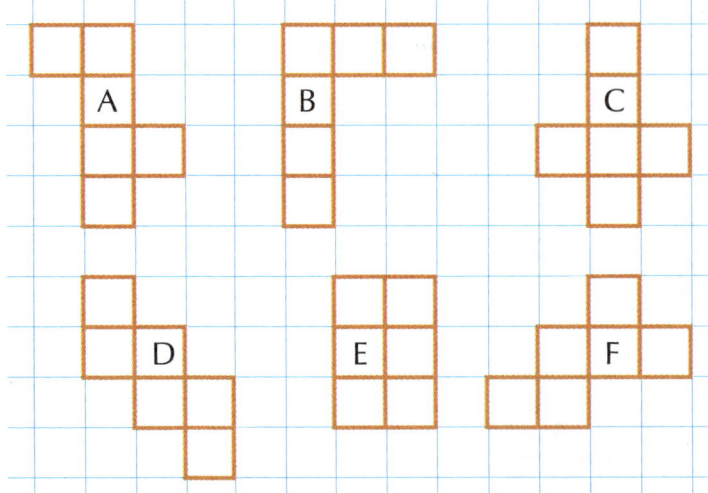

Hint: if you are not sure, cut out copies on squared paper to check.

Extension

6 · 1 Zelda drew this net of a pyramid.

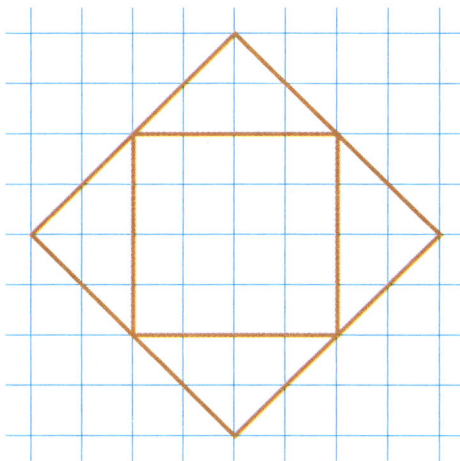

Lucinda said it will not work. Why not?

_____

Hint: if you are not sure, try copying it and cutting it out.

6 · 2 Here is a net for a dice.

On a dice, 6 is opposite 1, 5 is opposite 2, and 4 is opposite 3.

Mark 4, 5 and 6 on this dice.

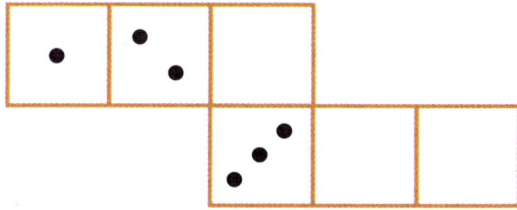

Hint: if you are not sure, cut out a copy of the net and fold it up.

J · 3 Is it possible to draw a net for a cone? If so, what will it look like? If not, explain why not.

_____

_____

3.3 Map scales

In this section you will learn how to:
● find distances on a map.

Key words
scale

WORKED EXAMPLE

These two cities are marked on a map with a scale of 1 cm to 20 km.

● Newingham

Birmcastle ●

How far apart are they?

Solution

First measure the distance between them. ——————→ It is 12 cm.

The scale tells us that each centimetre represents 20 km on the ground.

The actual distance is 12 × 20 = 240 km.

EXERCISE 3C

1 These two villages are on a map with a scale of 1 cm to 1 mile.

Axton ●   How far apart are the villages?

_____ miles

● Coldford

2 The scale of this map is 1 cm to 2 km.

windmill ●   ● tower

● bridge

church ●

Find the distances between the following.

a The church and the windmill. _____ km

b The windmill and the tower. _____ km

c The tower and the bridge. _____ km

3 Power lines are to be laid from a power station to the points marked *A*, *B* and *C*.

• *C*

B •

• power station

A •

a Draw the lines of the cables on the map.

b The scale of the map is 1 cm to 10 miles. Find the total length of the three cables.

_____

S 4 A map has a scale of 2 cm to 1 km.

You have been asked to write a table to convert map distances to distances on the ground. Put the missing numbers in this table. Add some more of your own.

On the map in cm	2 cm	10 cm				
On the ground in km						

5 A map has a scale of 2 cm to 1 km.
Here are some distances between places on the ground.
How far apart will they be on the map?

a 8 km _____   **b** 6.5 km _____   **c** 4.3 km _____   **d** 7.2 km _____

6 A road map has a scale of 1 cm to 20 km.
This scale can be used to convert map distances into distances on the ground.

a Fill in the missing kilometres.

b Convert these distances on the map into kilometres on the ground.

i 3 cm _____   **ii** 4.5 cm _____   **iii** 8.2 cm _____   **iv** 5.8 cm _____

Extension

PS 1 Here is a castle marked on a map. The scale is 1 cm to 2 km.

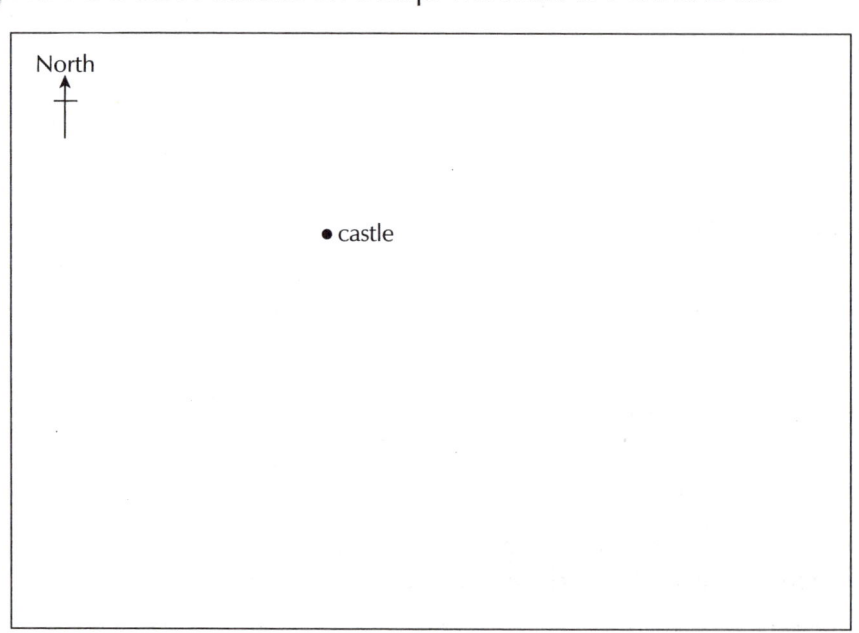

A station is 12 km east of the castle.

A café is 9 km south of the castle.

a Mark the station and the café on the map.

b How far apart are the station and the café?

_____ km

J 2 Pete has two maps of the same location.
The first has a scale of 1 cm to 1 mile.
The second has a scale of 1 cm to 2 miles.
On the first map, two villages are 14 cm apart.

How far apart will they be on the second map?

Explain how you worked this out.

_____

_____

3.4 Using scales

In this section you will learn how to:
● use scales and interpret drawings in different contexts.

Key words
net
scale

EXERCISE 3D

M 1 Ordnance Survey maps of the countryside use a scale of 1 to 25 000. They are very popular with walkers.

This means that 1 cm on the map represents $\frac{1}{4}$ of a kilometre.

a How many centimetres on the map represent 1 km? _____

b How many centimetres on the map represent 3 km? _____

c The distance between two hill tops on the map is 10 cm.
How far apart are the hill tops? _____

d Alan is planning a walk. He estimates the distance on the map is 34 cm. How long is the walk? _____

e These maps have a square grid printed on them. The grid lines are 4 cm apart.
How will these help you judge distances when you are reading the map?

_____

PS 2 The standard size for a shoe box from one famous shoe manufacturer is 32 cm by 17 cm by 12.5 cm.

Here is a sketch of the net for a shoe box. It does not have a lid.

Write the lengths on as many lines in this net as you can.

AU 3 At the time of writing, the world triple jump record for men is 18.29 metres. Make a sketch with a man standing next to the length he completed. Say how you did it.

_____

_____

4 Roads in most European countries have speed limits given in km/h and not in mph. Car speedometers in the UK show both scales.

a This table shows speed limits in France in km/h. Use the speedometer to help you put in the corresponding speeds in mph.

	Motorway	Motorway when wet	Open road	Open road when wet	Dual carriageway	Dual carriageway when wet	Town
km/h	130	110	90	80	110	100	50
mph							

b How are speed limits in France different from speed limits in England?

_____

_____

Extension

1 Here are the heights of four famous structures.

Statue of Liberty in New York	92 metres
Eiffel Tower in Paris	300 metres
Leaning Tower of Pisa in Italy	55 metres
Great Pyramid of Giza in Egypt	147 metres

Hint: roughly how many times bigger is the Eiffel Tower than the Statue of Liberty?

On a separate piece of paper, draw a sketch of each, side by side, to show their relative sizes.

checklist

☐ I can read scales and estimate distances.

☐ I can recognise nets of different 3D objects.

☐ I can find distances on a map.

☐ I can use scales and interpret drawings in different contexts.

You could work with a partner on this task.

A friend has asked you to help with the design for the back garden of a new house.

You need to include the following items in the design.

- A patio near the house, big enough for a table and chairs.
- A circular pond with a diameter of 2 metres.
- A section to grow vegetables that is at least 12 m^2 in area.
- A flower bed that is no wider than 1 metre.
- An area of grass with a children's slide on it.
- Paths that are 0.5 metres wide in suitable places.

The plot of land is 7 m wide and 10 m long.

The house is next to the shorter side.

Task

1 Make a scale drawing of a plan for the garden.

a Start by working out the area of the piece of land.

b Now draw a rectangle to represent the piece of land. Use a scale of 2 centimetres to 1 metre.

c Now you can start your design. Make sure your design includes all the features in the white box.

   You will need to make some decisions for yourself. For example:

   – Where is the best place to put the vegetable plot?

   – How big does the patio need to be?

   – How much room does the slide need?

   – Where should any paths go?

   Include any relevant measurements on your plan.

2 When you have finished, compare your plan with someone else's. What are the similarities? What are the differences?

4 Number: Speed and proportion

4.1 Speed

In this section you will learn how to:
- make calculations about speed.

Key words
distance
kilometres per hour (km/h)
miles per hour (mph)
speed

WORKED EXAMPLE

Mr Wheeler drove for 2 hours at an average speed of 35 miles per hour (mph) and then for 3 hours at an average speed of 30 mph.

What was his average speed for the whole journey?

<u>Solution</u>

First we need to find the distance travelled.

2 hours at 35 mph is 2 × 35 = 70 miles.

3 hours at 30 mph is 3 × 30 = 90 miles.

Total distance = 70 + 90 = 160 miles.

Total time = 2 + 3 = 5 hours.

Average speed = 160 ÷ 5 = 32mph. $\longrightarrow$ Remember: speed = $\dfrac{\text{distance}}{\text{time}}$

EXERCISE 4A

1 Geri cycles at an average speed of 12 mph. How far will she go in these times?

a 1 hour _____ **b** 2 hours _____ **c** 5 hours _____

FM Functional Maths **AU** (AO2) Assessing Understanding **PS** (AO3) Problem Solving

2 A racing cyclist can average 25 mph. How far will she travel in these times?

    a 2 hours _____   **b** 4 hours _____   **c** 6 hours _____

3 A fast train can travel at 90 mph. How far can it travel in these times?

    a 1 hour _____   **b** 2 hours _____   **c** 4 hours _____

4 30 mph is the speed limit in built-up areas.

Complete this table to show how far you can travel at that speed in different times.

Time (hours)	0.5	1	2	3	4
Distance (miles)		30			

5 A garden snail has been recorded as travelling at a speed of 17 cm per minute.

How far can it go in these times?

    a 2 minutes _____   **b** 4 minutes _____   **c** 10 minutes _____

6 A car travels 60 miles in 2 hours. What is its average speed? _____ mph

7 Find the average speeds for these journeys by bicycle.

	Distance	Time	Average speed in miles per hour
a	14 miles	2 hours	
b	45 miles	3 hours	
c	48 miles	4 hours	

8 Bruce can walk at 4 mph.

    a How far can he walk in 3 hours? _____ miles

    b How long will he take to walk 20 miles? _____ hours

9 The speed limit on a motorway is 70 mph.

At that speed, how far can you drive in these times?

a 1 hour _____

b 3 hours _____

c $\frac{1}{2}$ hour _____

d $1\frac{1}{2}$ hours _____

10 The distance from Exeter to Carlisle is about 400 miles.

How long will that take at an average speed of 50 mph? _____

11 In France, Marie drives 120 kilometres in 2 hours.

What is her average speed in kilometres per hour (km/h)? _____

12 Karl cycles at an average speed of 24 km/h.

How far will he travel in these times?

a 1 hour _____ km

b 2 hours _____ km

c 30 minutes _____ km

d 15 minutes _____ km

Extension

1 Fran cycled from Norwich to Great Yarmouth in 2 hours and then back in 3 hours. The route she chose was 30 miles each way.

a What was her average speed for each part of the journey?

Going: _____   Coming back: _____

b What was her average speed for the whole journey?

Hint: you need to know the total distance and the total time.

_____

2 Helga cycled 45 km in 2 hours and 30 minutes.

What was her average speed in km/h ? _____

4.2 Using proportion in different contexts

In this section you will learn how to:
● use the idea of proportion in different contexts.

Key words
best buys
exchange rate
speed

EXERCISE 4B

1 Here is a list of ingredients needed to make eight brownies.

a Write down the amount of each ingredient you will need to make 16 brownies.

Butter: _____ g   Cocoa: _____ g

Chocolate: _____ g   Sugar: _____ g

Flour: _____ g   Eggs: _____

Ingredients for Brownies	
Butter	90 g
Chocolate	120 g
Flour	30 g
Cocoa	20 g
Sugar	150 g
Eggs	2
Makes 8 brownies	

b Max only has three eggs.
How many brownies can he make?

_____

c How much butter would you need for 24 brownies?

_____

2 The exchange rate is 1.45 dollars to the pound.

Hint: this means that with £1 you can buy 1.45 dollars – that is, 1 dollar and 45 cents.

a How many dollars can you buy for £10 ? _____

Hint: you will need to do a division. Use a calculator!

b How many dollars can you buy for £200 ? _____

c How many pounds can you get for 500 dollars ? _____

 5 miles is approximately the same as 8 kilometres.

a How many kilometres are 10 miles ? _____

b How many kilometres are 20 miles ? _____

c How many miles are 24 kilometres ? _____

d Fill in the gaps in this table.

miles	5	50		100	200	
kilometres	8		96			400

e Erin said that a mile is more than 1 kilometre but less than 2 kilometres. Do you agree? Explain your answer.

_____

f The circumference of the Earth is about 25 000 miles. What is that in kilometres? Circle the correct answer from this list.

400 km   4 000 km   40 000 km   400 000 km   4 000 000 km

Extension

FM 1 A cereal comes in two pack sizes:
A small (200 g) pack costs £1.88
A large (300 g) pack costs £2.55

a What is the cost of 100 g in the **small** pack?

_____

b What is the cost of 100 g in the **large** pack?

_____

c Which size is better value? _____

Hint: compare your answers to parts a and b.

M **2** The sme type of crystal glasses is sold in two different packs.

Small pack	Large pack
Contents	Contents
4 glasses	**12 glasses**
£3.20	£10.20

Which size is the better value for money?

You **must** show your working.

_____

(Total 2 marks)

AQA, November 2005, Module 3 Foundation, Question 17

3 The ingredients needed to make 500 millilitres (ml) of a fruit drink are

orange juice	300 ml
mango juice	60 ml
lemonade	140 ml

a What percentage of the fruit drink is orange juice? _____ % (2)

b Robert wants to make 750 ml of the fruit drink.

How much lemonade will he need? _____ ml (2)

(Total 4 marks)

AQA, November 2006, Module 3 Foundation, Question 18

checklist

☐ I can make calculations about speed.

☐ I can use the idea of proportion in different contexts.

You have invited three friends around for dinner and chosen recipes for a starter, a main course and a dessert which you would like to cook for them. However, the recipe ingredients are not written for four people. Re-write them for four people.

Getting started

Here are some things to bear in mind.

- You should round off quantities sensibly if necessary. For example, you might want $\frac{1}{2}$ of 85 g which is 42.5 g but that would be silly in a recipe. 40 g or 45 g would be better.

- Sometimes you may need to adjust things. For example, you cannot include fractions of an egg. However, they do come in different sizes.

- Recipes can be flexible. A little bit more or less of an ingredient will not usually make much difference.

Starter

Indian potato cakes
Serves 8

650 g potatoes
40 g plain flour
115 g peas
115 g carrots
1 onion
1 red chilli
1 clove of garlic
1 teaspoon of cinnamon
1 teaspoon of cumin
Juice of $\frac{1}{2}$ a large lime
2 tablespoons of chopped coriander leaves
Salt and oil

Main
Aubergine meat balls
Serves 6

900 g aubergines
2 onions
450 g minced lamb
50 g grated Parmesan cheese
2 large eggs
Salt and pepper

Dessert
Apple and cinnamon pancakes
Serves 3

60 g plain flour
30 g wholemeal flour
1 large egg
150 ml of milk and water mixed
$\frac{1}{4}$ teaspoon of vanilla essence
$\frac{1}{2}$ teaspoon of ground cinnamon
2 small Cox's apples
3 tablespoons of dry cider

5 Geometry: Area and perimeter

5.1 Perimeter

In this section you will learn how to:
- find and use the perimeter of a shape.

Key words
perimeter
rectangle
square

WORKED EXAMPLE

Here is a sketch of a field.

The perimeter of the field is 205 metres.
The longest side is 75 metres. The opposite side is 40 metres.
The other two sides are the same length.

How long are the other two sides?

Solution

40 + 75 + third + fourth = 205 ⟶ The perimeter is the distance all around the field.

115 + third + fourth = 205 ⟶ 40 + 75 = 115

Third + fourth = 205 − 115 = 90

The other two sides are both 45 metres. ⟶ Half of 90 is 45.

EXERCISE 5A

1 Work out the perimeters of the following shapes.

a
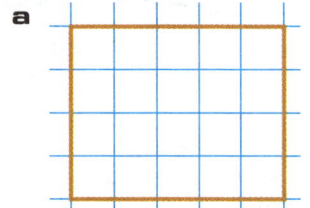

b
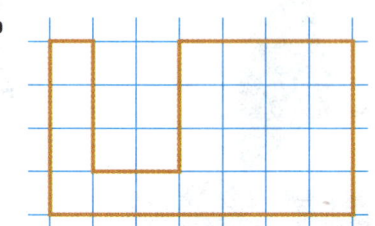

Each square on the grid is 1 cm². Count the squares to find the area.

a Perimeter = _____ cm   **b** Perimeter = _____ cm

FM Functional Maths **AU** (AO2) Assessing Understanding **PS** (AO3) Problem Solving

2 Find the perimeters of these fields.

a

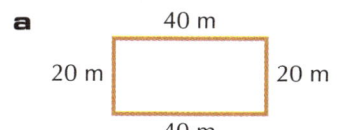

Perimeter = _____ m

b

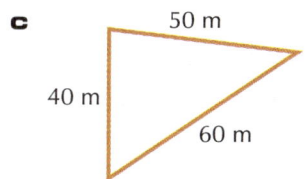

Perimeter = _____ m

c

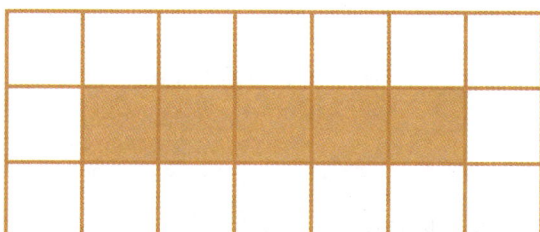

Perimeter = _____ m

U 3 The grids in this question are made from centimetre squares.

Work out the perimeter of the shaded rectangle.

_____ cm   (1)

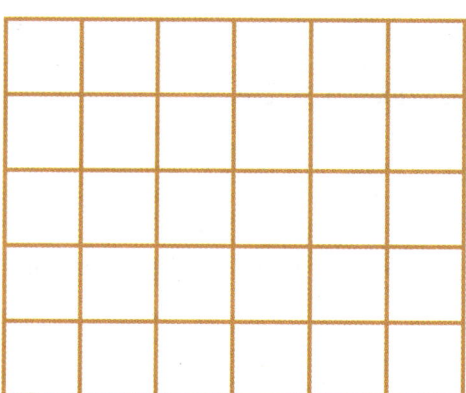

On the grid, draw and shade a different rectangle which has the same perimeter.   (2)

(Total 3 marks)

AQA, November 2008, Module 5 Foundation, Question 1

4 Calculate the perimeters of these rectangles.

Hint: write in the two missing lengths.

a

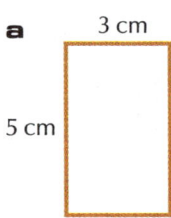

3 cm

5 cm

a Perimeter = _____

b

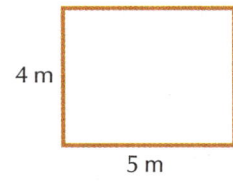

4 m

5 m

b Perimeter = _____

c

17 mm

1 mm

c Perimeter = _____

AU **5** The diagram shows a pentagon.

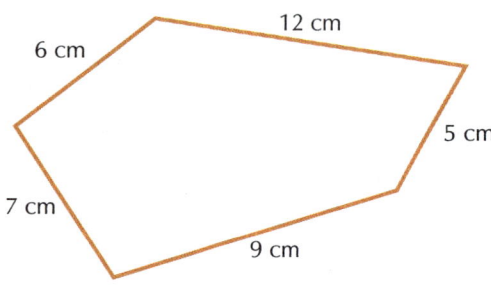

12 cm

6 cm

5 cm

7 cm

9 cm

Not drawn accurately

a Explain how you can show that the perimeter is equal to 39 cm.

_____ (1)

b Work out the difference between the length of the longest side and the length of the shortest side.

_____ (2)

c The diagram shows a regular hexagon.

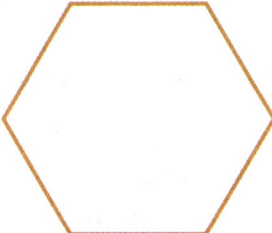

Not drawn accurately

The perimeter of the hexagon is equal to the perimeter of the pentagon.

Work out the length of one side of the hexagon.

_____ (2)

(Total 5 marks)

AQA, June 2007, Module 5 Foundation, Question 5

M 6 Mr Williams wants to put a small fence all the way around his vegetable plot.
A diagram of the plot is shown below.

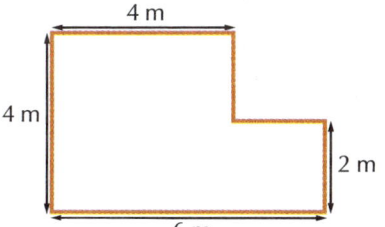

What is the total length of fence he needs?

_____

Extension

U 1 An equilateral triangle and a square both have a perimeter of 36 cm.

On a separate piece of paper, sketch each shape and write the lengths on each side.

S 2 Each small square has a perimeter of 20 cm.

What is the perimeter of the large square?

Hint: find the length
of each small square.

_____ cm

5.2 Area

In this section you will learn how to:

● calculate the area of rectangles and more complicated shapes.

Key words

area
rectangle
square

WORKED EXAMPLE

Here is a sketch of a garden lawn.

Find the area of the lawn.

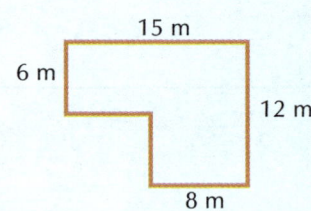

Solution

First divide the shape into rectangles.

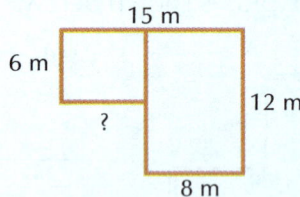

The area of the larger rectangle is $8 \times 12 = 96$ m^2. → Don't forget the units.

The length marked with a ? is $15 - 8 = 7$. → ? + 8 makes 15.

The area of the smaller rectangle is $6 \times 7 = 42$ m^2. →

The total area of the lawn is $96 + 42 = 138$ m^2. Add the two areas.

EXERCISE 5B

1 Work out the areas of the following shapes.

a

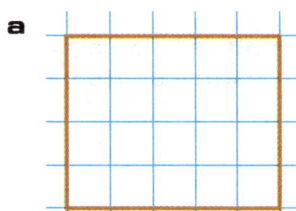

b

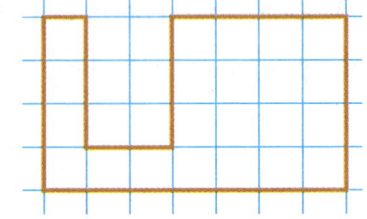

a Area = _____

b Area = _____

2 The map shows a small island.
Each square on the map represents 1 km².
Estimate the area of the island.

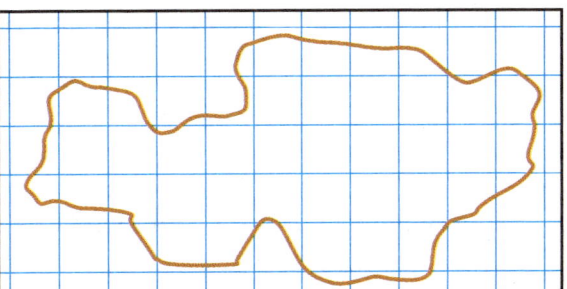

Area = _____

AU 3 A triangle and a hexagon are made
from identical equilateral triangles.

 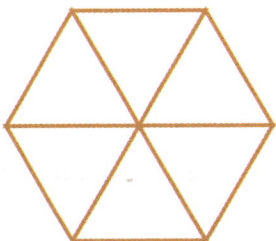

a Do the triangle and hexagon have the same perimeter?

Tick the correct box.   Yes [ ]   No [ ]

Explain your answer.

_____ (1)

b Do the triangle and hexagon have the same area?

Tick the correct box.   Yes [ ]   No [ ]

Explain your answer.

_____ (1)

(Total 2 marks)

AQA, June 2009, Paper 2 Foundation, Question 6

4 Calculate the areas of these rectangles.

> **Don't forget** the units in your answers.

a 3 cm   **b**

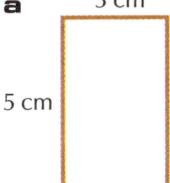

5 cm

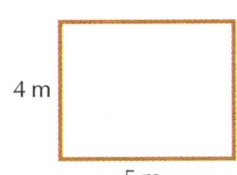

4 m

5 m

c

17 mm

1 mm

a Area = _____   **b** Area = _____   **c** Area = _____

5 The diagram shows the measurements of a rectangle.

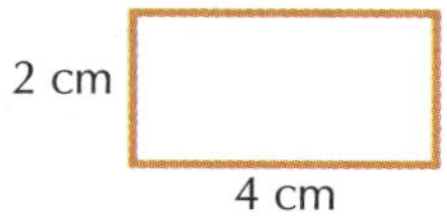

2 cm ⬚ Diagram not drawn accurately

4 cm

Four of the rectangles are arranged to form a larger rectangle.

Diagram not drawn accurately

a Work out the perimeter of the larger rectangle.

_____ cm (2)

b Work out the area of the larger rectangle.

_____ cm² (2)

(Total 4 marks)

AQA, November 2008, Paper 1 Foundation, Question 11

6 Here is a plan of a vegetable plot.

The plot is divided into two rectangles.

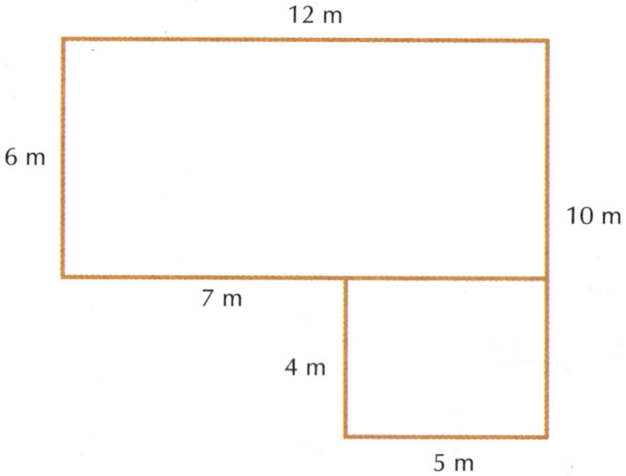

a Find the area of each rectangle.

12 m

6 m

10 m

7 m

4 m

5 m

_____

b Find the area of the vegetable plot.

_____

Extension

FM 1 Jack wants to carpet this floor. He needs to find the area of the floor.
Calculate the area for him.
Show your method.

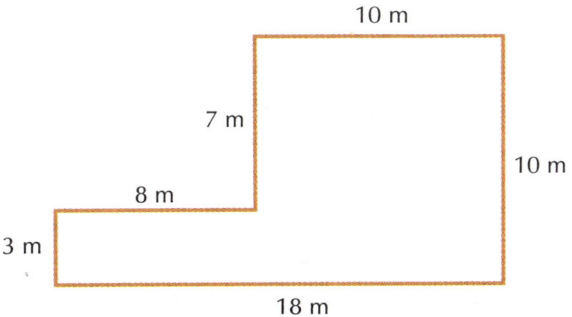

_____

2 a Find the area of the rectangle.

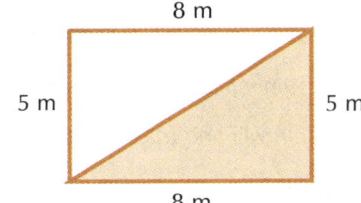

_____

b The triangle is half the size of the rectangle.

Find the area of the triangle.

_____

3 Find the areas of these triangles.

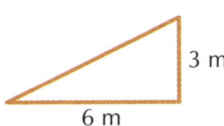

a _____   **b** _____

5.3 Using perimeter and area

In this section you will learn how to:
- calculate perimeter and area in practical situations.

Key words
area
perimeter

WORKED EXAMPLE

Here is a sketch of a garden lawn.

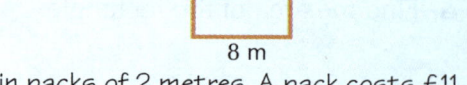

Anish wants to lay new turf and put edging all round the lawn.

Turf costs £3.40 per square metre.

Edging is needed for the border of the lawn. Edging comes in packs of 2 metres. A pack costs £11.

How much will the materials cost?

Solution

To find the cost of the **turf**, we need to know the **area** of the lawn.

We did that in the example in topic 5.2.

The area is 138 m². ——————————→ Look back at the example in 5.2 to see how to do that.

The turf will cost 138 × £3.40 = £469.20.

To find the cost of the **edging** we need to know the **perimeter** of the lawn.

That is 15 + 12 + 8 + 6 + 7 + 6 = 54 m. ————→ Can you see where those numbers come from?

The edging is £11 for 2 metres.

54 ÷ 2 = 27 packs needed.

The edging will cost 27 × £11 = £291.

Total cost is £469.20 + £291 = **£760.20**. ———→ Anish may want to order a little more to allow for wastage. He may also need to pay for delivery.

EXERCISE 5C

You will need a calculator for this exercise.

The dimensions of many games courts and pitches were fixed years ago. They were given as feet or yards. When these are changed into metres, they become awkward decimals.

PS 1 A tennis court is 78 feet long (that is, 23.77 metres).

A singles tennis court is 27 feet wide (that is, 8.23 metres).

a Find the perimeter of a singles tennis court in feet.

_____ feet

b Find the area of a singles tennis court in square feet.

_____ square feet

PS 2 A doubles tennis court is 78 feet long and 36 feet wide.

Find the perimeter and area of a doubles tennis court.

Hint: remember the units.

a Perimeter = _____ feet

b Area = _____ square feet

AU 3 Is a doubles tennis court twice as big as a singles court? Give a reason for your answer.

_____

AU 4 All badminton courts are 44 feet long.

A singles court is 17 feet wide and a doubles court is 20 feet wide.

Find the perimeters and areas of a singles court and a doubles court.

Put units in your answers.

Hint: the units will be feet or square feet. Draw diagrams to help you on another piece of paper.

a Singles court perimeter = _____

Singles court area = _____

b Doubles court perimeter = _____

Doubles court area = _____

FM **5** A hockey pitch is 60 yards (55 metres) long and 100 yards (91.4 metres) wide.

a The players run around the pitch to warm up. How many yards is that?

> **Hint:** draw a diagram.

_____

b The surface of the pitch needs to be replaced. The cost depends on the area. What is the area of a hockey pitch in square yards?

_____

FM **6** Mike wants to make a new lawn in his garden.

The dimensions are shown on this diagram.

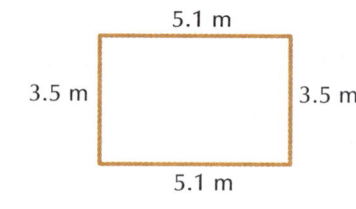

a Mike wants to grow the lawn from seed.

To do that, does he need to know the perimeter or the area of the lawn? _____

b Use a calculator to find both the perimeter and the area of Mike's lawn.

Perimeter = _____   Area = _____

> **Hint:** what might you put round a lawn?

c Why might Mike want to know the perimeter of the lawn?

_____

FM **7** This is the floor of a room.

a Jude wants to buy a carpet that costs £8.49 per square metre.

What will the carpet for this room cost? Show your method.

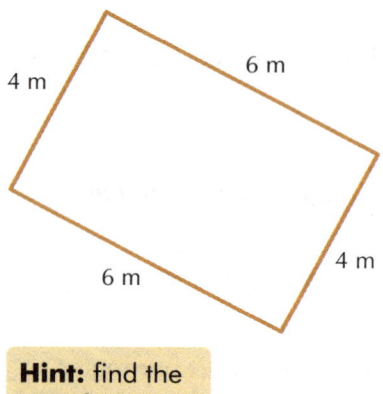

> **Hint:** find the area first.

£_____

b Skirting board goes round the edge of the room.

It costs £20 for a piece of skirting board 2.5 metres long.

How much will it cost to put board around the room?

Show how you got your answer.

> **Hint:** find the perimeter first.

£_____

c You will not need skirting board in front of the door.
Will this make any difference to the cost? Explain your answer.

_____

Extension

FM 1 Mr Waters want to put carpet tiles in his office.

The tiles are 50 cm squares.

They cost £3.49 each.

Here is a plan of his office floor.

4.8 m

3.3 m   3.3 m

4.8 m

He has asked for your advice about the number of tiles he needs and the cost of buying them.

What can you tell him?

_____

checklist

☐ I can find and use the perimeter of a shape.

☐ I can calculate the area of rectangles and more complicated shapes.

☐ I can calculate perimeter.

You could work with a partner on these tasks.

Surprisingly, there is no fixed size for a football pitch.

The FA suggests the following maximum and minimum lengths and widths.

	Length in metres		Width in metres	
	maximum	minimum	maximum	minimum
Seniors	120	90	90	45.5
Under 16	100.6	82.3	64	45.5
Under 14	91	72.8	56	45.5
Under 12	82	68.25	50.77	42

120 m by 90 m

Task A

1 The table shows that the **maximum** size for a **senior** pitch is 120 metres long and 90 metres wide.

   What are the **maximum** area and perimeter of a senior pitch?

2a What is the **minimum** size for a **senior** pitch?

 b What would the area and perimeter of this pitch be?

3 Look at the difference between your answers to questions 1 and 2.

   Would you say there is a **big difference** or a **small difference** between them?

   Discuss your answer with someone else and see if they agree.

Task B

1 Make a **scale drawing** to illustrate the maximum-sized and minimum-sized senior pitches that you looked at in questions 1 and 2 of Task A.

2 The table shows the maximum and minimum suggested sizes for three different ages.

   Choose **one** of these ages and compare the maximum-sized and minimum-sized pitches.

   Hint: do it in the same way as you did for the seniors.

3 Try to find out the size of the football pitches in your school.

   You could do this by measuring them.

   How do they compare with the suggested sizes in the table?

4 The council are going to provide a football pitch near where you live for the community to use. Recommend a size and give a reason for your answer.

Rory Delap
Stoke City – midfield

Interesting fact:

The Premier League has its own regulations about pitch sizes.

When Stoke City were promoted to the Premier League in 2008, they opted to use the smallest size pitch allowed. Some people say they might have done this to help their player Rory Delap, who has one of the longest throw-ins in football.

6 Geometry: Symmetry

In this section you will learn how to:
- reflect a shape in a mirror line
- find lines of symmetry.

Key words
reflection
symmetry

WORKED EXAMPLE

How many lines of symmetry do a square and a parallelogram have?

<u>Solution</u>

Remember that if we fold a shape along a line of symmetry, the two parts will exactly match.

We can do that in four ways with a square.

A square has **four** lines of symmetry. ⟶ You can check by folding a square of paper.

This is a parallelogram.

It has **no** lines of symmetry. ⟶ Do you disagree? Check by folding a parallelogram of paper.

EXERCISE 6A

1 Draw the lines of symmetry on the following shapes.

a

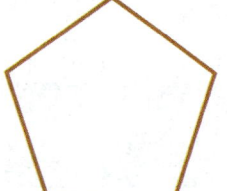

b

c

FM Functional Maths **AU** (AO2) Assessing Understanding **PS** (AO3) Problem Solving

d    **e** 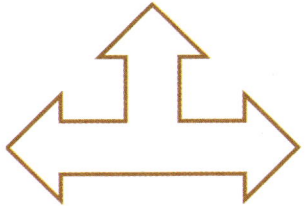   **f**

2 How many lines of symmetry do these icons have?

a 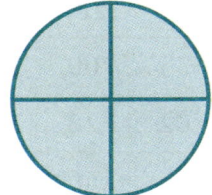   **b**    **c**

_____   _____   _____

3 Complete the shapes on the grid so that the black line is a line of symmetry.

a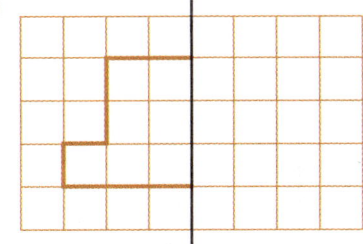

b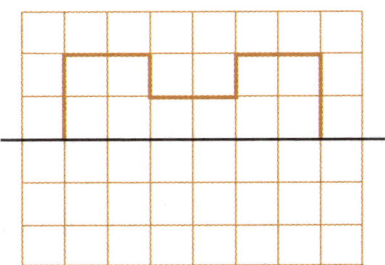

4 Draw the reflection of these shapes to make the blue and black lines into lines of symmetry and name the complete shapes.

a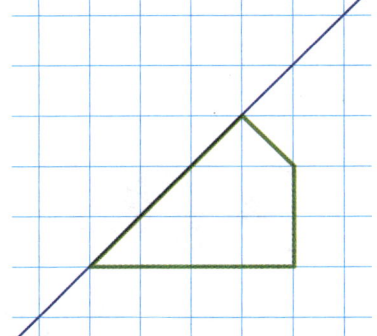

Name of the complete shape:

_____

b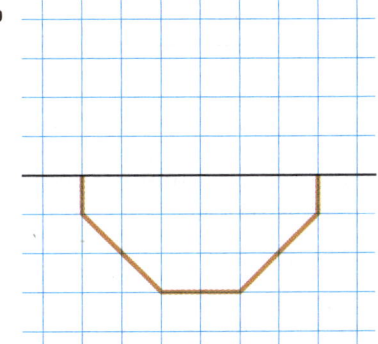

Name of the complete shape:

_____

5 Which of these letters have:

a no lines of symmetry? _____

b exactly one line of symmetry? _____

c two lines of symmetry? _____

SHAPE

6 This multiplication table is **symmetrical**.

Use the **symmetry** to fill in the missing numbers **without doing any calculations**.

Hint: the answers are right in front of you!

×	25	26	27	28	29
25	625	650	675	700	725
26	650	676	702	728	754
27	675		729	756	783
28	700			784	812
29	725				841

Extension

1 a Reflect the orange shape in the blue line.

b Reflect **both** parts in the black line.

c How many lines of symmetry does the finished shape have?

_____

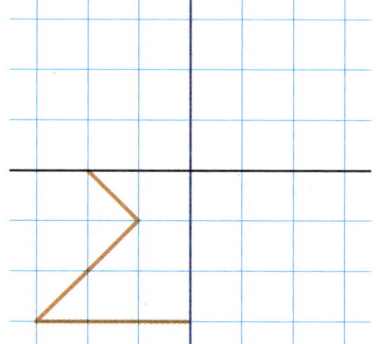

2 a Draw a horizontal line of symmetry through this word.

b Write another word with a horizontal line of symmetry.

_____

CODE

6.2 Rotation symmetry

In this section you will learn how to:
● recognise and describe rotational symmetry.

Key words
order
rotational symmetry

WORKED EXAMPLE

Describe the symmetry of this shape.

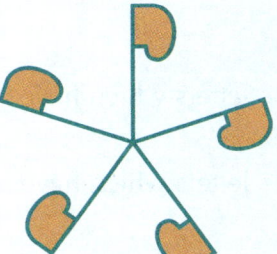

Solution

We can look for reflections or rotations.

It has no lines of symmetry.

It has rotational symmetry of order 5. ⟶ Check with a piece of tracing paper.

If you turn it around, it will match up five times in one whole turn.

EXERCISE 6B

1 What is the order of rotation of the following shapes?

Hint: shape **g** has order 1. We say it has no rotational symmetry.

a

order: _____

b

order: _____

c
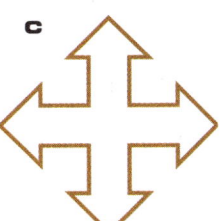
order: _____

d

order: _____

e

order: _____

f
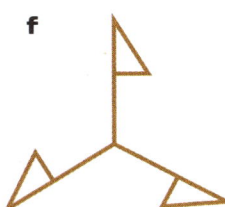
order: _____

g

order: _____

h

order: _____

2 This shape has no rotational symmetry. The order of rotational symmetry is 1. Continue this drawing to make it have a rotational symmetry of order 4.

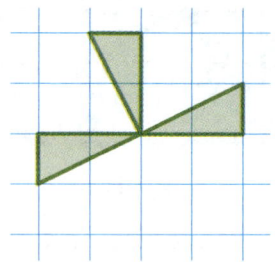

3 Here are five letters.

FUMEX

    a Write down the letters which have no lines of symmetry. _____ (2)

    b Write down the letters which have rotational symmetry. _____ (2)

                                                          (Total 4 marks)

AQA, November 2008, Module 5 Foundation, Question 6

4 a Draw the lines of symmetry on these shapes.

                                                                     (3)

    b The shape below is made from squares.

        Add some more squares to form a shape with rotational symmetry of order 2.

                                                                         (1)

                                                          (Total 4 marks)

AQA, June 2008, Paper 2 Foundation, Question 8

5 For each of these letters, give the order of rotational symmetry and the number of lines of symmetry.

	Order of rotational symmetry	Lines of symmetry
S		
N		
W		
X		
Z		

Extension

AU 1 This word has rotational symmetry of order 2 but no lines of symmetry.

N O O N

a Find other words with rotational symmetry and no line symmetry and words with line symmetry but no rotational symmetry.

_____

b Do you know any words which have both rotational and line symmetry?

_____

c Compare your results with other people's.

Hint: start with letters that look the same when you turn them upside down.

PS 2 Draw a picture that has:

• rotational symmetry of order 5

• no lines of symmetry.

Use this pentagon as a starting point.

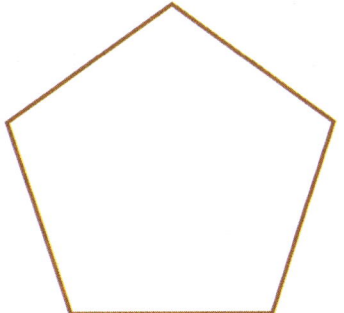

6.3 Symmetry in everyday situations

In this section you will learn how to:
● recognise different types of symmetry in realistic situations.

Key words
line of symmetry
order of rotational
symmetry

EXERCISE 6C

1 You have probably seen this tangram puzzle before.

They have been used for thousands of years to help people practise their spatial understanding.
Seven pieces fit together to make a square.

Trace this shape onto a piece of paper and cut out the pieces.

a Which piece has four lines of symmetry?

_____

b Which piece has no lines of symmetry? _____

c Which **two** pieces have rotational symmetry? _____

d Use the cut out pieces to make another shape with rotational symmetry.

e Now use them to make a shape with line symmetry.

2 Here are some road signs

Draw in any lines of symmetry.

If it has rotational symmetry, write the order of rotational symmetry under it; otherwise write 'none'.

a

b

c

_____   _____   _____

d

e

f

_____   _____   _____

3 Here are two car logos. Do you recognise them?

a For each one find the number of lines of symmetry and the order of rotational symmetry (if any).

i

ii

_____   _____

_____   _____

b See if you can find any other car logos which have rotational or line symmetry. Sketch them here and describe the symmetry.

_____

_____

Extension

1 This is a tile.

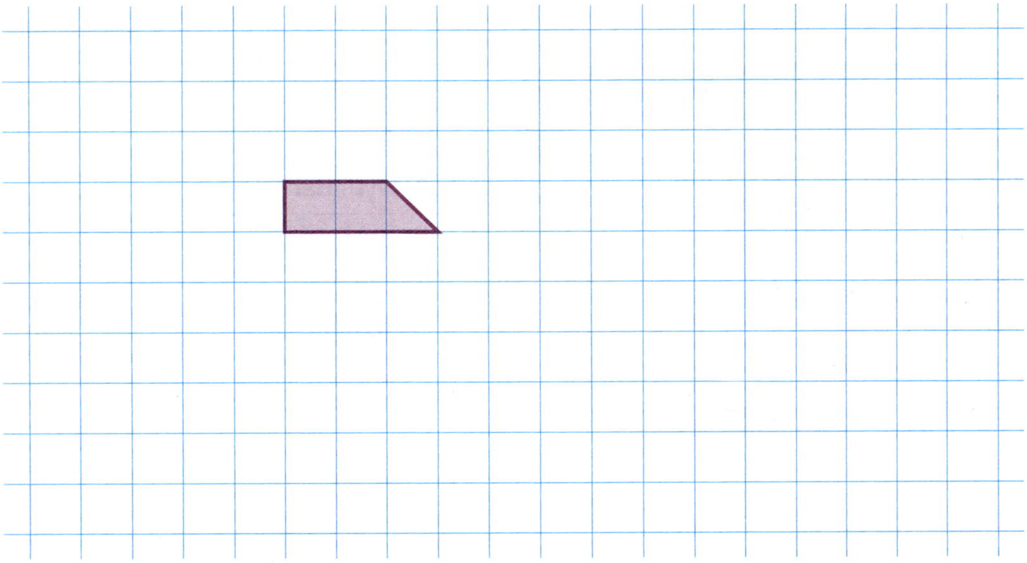

Use tiles of this shape to make a symmetrical pattern.

Describe the symmetry of your pattern.

_____

_____

Hint: does your pattern have lines of symmetry or rotational symmetry?

☐ I can reflect a shape in a mirror line.

☐ I can find lines of symmetry.

☐ I can recognise and describe rotational symmetry.

☐ I can recognise different types of symmetry in realistic situations.

You are a designer working for a leading car manufacturer. You have been set the task of designing the hub cap for a new model.

Task A

Carry out market research into existing hubcaps by recording the symmetry of these hubcaps in the table.

All of the hubcaps have rotational symmetry to make the wheel balanced. Some of them also have lines of symmetry.

Task B

Design a new hubcap for a leading car manufacturer. It must have rotational symmetry to ensure balance.

Hubcap	Order of rotational symmetry	Lines of symmetry
1 Vauxhall	9	9
2 Ford		
3 Hyundai		
4 Ford		
5 Vauxhall		
6 Renault		
7 Citroën		
8 Almeira		
9 Saab		
10 Ford		
11 Audi		
12 Renault		
13 Smart		
14 Ford		
15 Fiat		

1 Vauxhall

2 Ford

3 Hyundai

4 Ford

5 Vauxhall

6 Renault

7 Citroën

8 Almeira

9 Saab

10 Ford

11 Audi

12 Renault

13 Smart

14 Ford

15 Fiat

7 Geometry and measures: Angles

7.1 Measuring angles

In this section you will learn how to:
- recognise, measure and draw angles.

Key words
acute
angle
obtuse
reflex

WORKED EXAMPLE

Measure the obtuse angle in this quadrilateral.

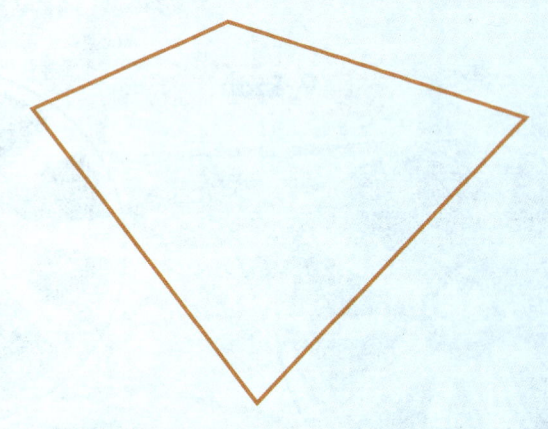

Solution

An obtuse angle is more than 90°.

In this shape it is the top one.

Place a protractor on the angle.

Choose the correct scale and read off the angle.

The obtuse angle is 138°.

> **Hint:** turn the book around, if it helps you to use the protractor.

An acute angle is less than 90°.

A reflex angle is more than 180°.

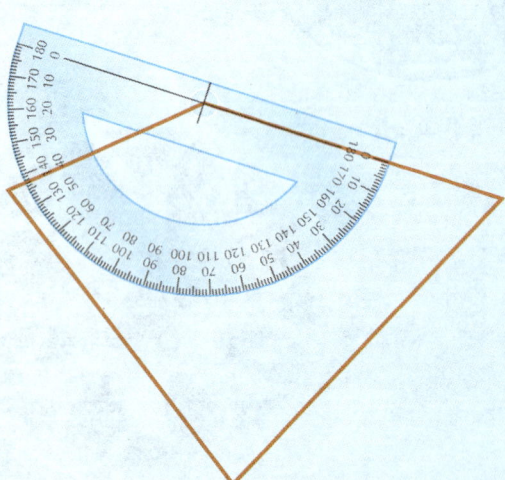

FM Functional Maths   **AU** (AO2) Assessing Understanding   **PS** (AO3) Problem Solving

EXERCISE 7A

1 Which of these are right angles? Circle the letters.

A   B   C   D

2 Here is a list of words connected with angles.

Acute   Full-turn   Obtuse   Reflex   Right   Straight

Choose the correct word to describe each of these angles.

a

Answer _____ angle

b

Answer _____ angle

c

Answer _____ angle

d

Answer _____ angle

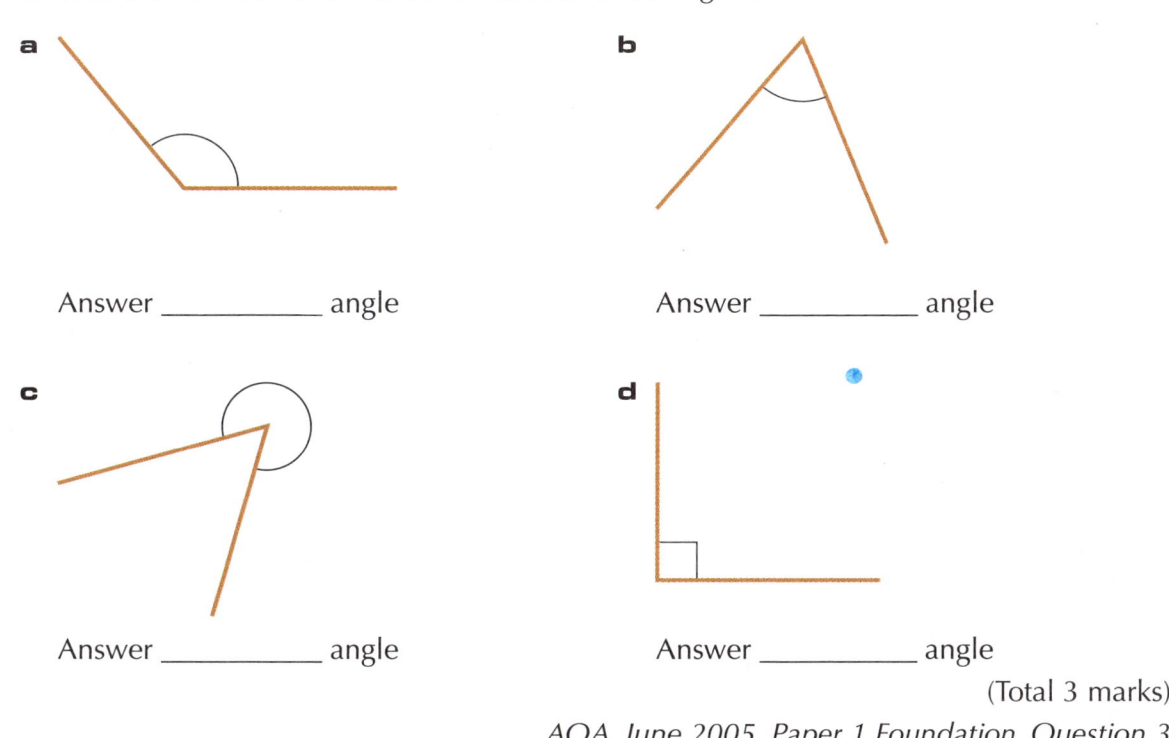

(Total 3 marks)

AQA, June 2005, Paper 1 Foundation, Question 3

3 Mark the obtuse angles inside this quadrilateral.

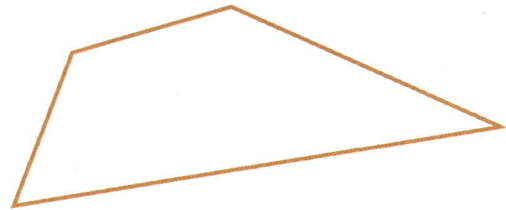

PS **4** List these angles in order of size, smallest first. _____

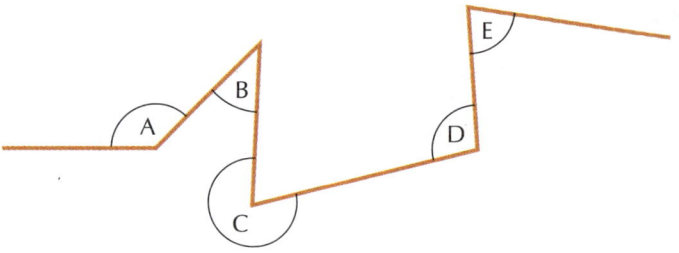

5 Measure these angles.

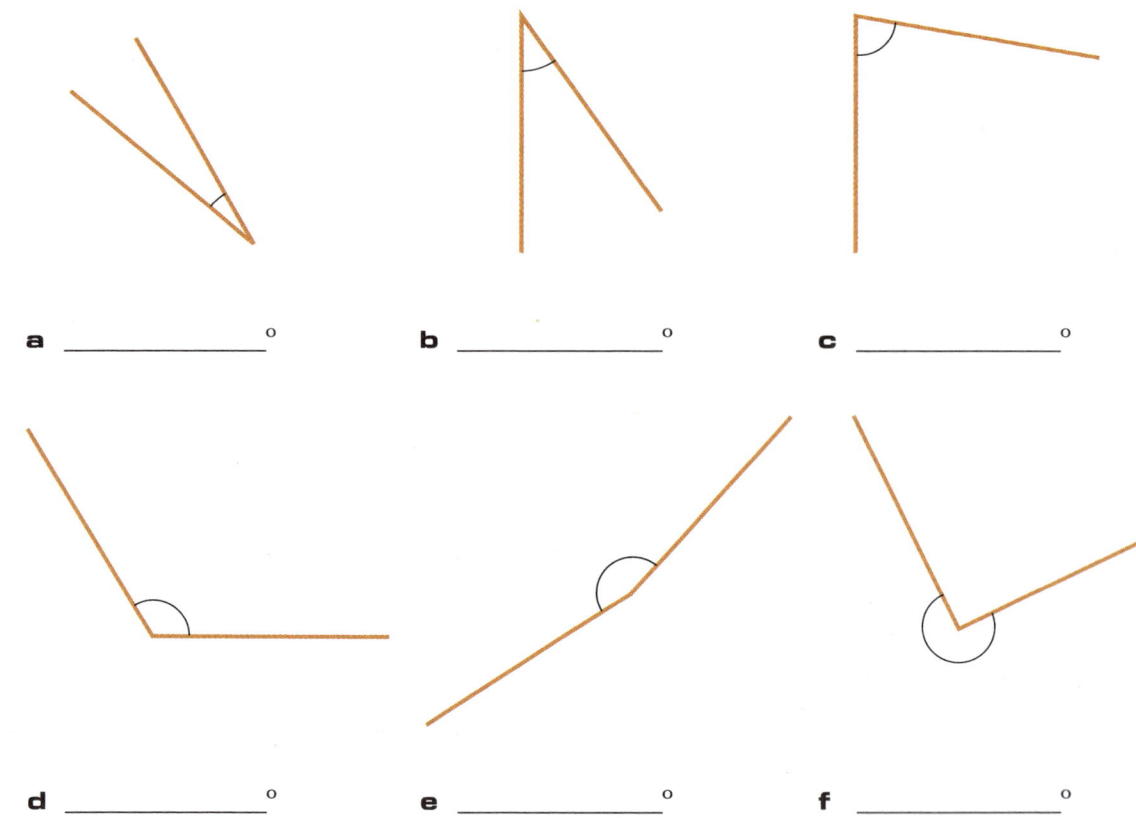

a _____ °

b _____ °

c _____ °

d _____ °

e _____ °

f _____ °

6 On one end of this line draw an angle of 45°.

On the other end draw an angle of 70°.

7 Draw angles of the following sizes.

    a 55°   **b** 12°

    c 120°   **d** 155°

Extension

U 1 This triangle has three acute angles.

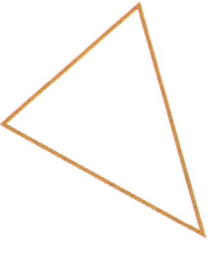

    a Draw a triangle that has one obtuse angle.

    b Can you draw a triangle with two obtuse angles? What happens when you try?

_____

_____

2 Draw a quadrilateral with one reflex angle.

7.2 Calculating angles

In this section you will learn how to:
● use known angles to calculate others.

Key words
angle
quadrilateral

WORKED EXAMPLE

Calculate angle *a* in this diagram.

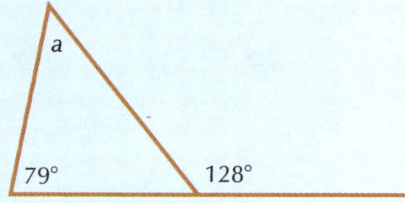

Solution

You must not try to measure it. The diagram is not accurately drawn.

First find the angle next to the 128° angle.

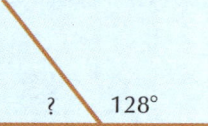

The angles on a straight line add up to 180 degrees. ⟶ Important fact.

? = 180° − 128° = 52°

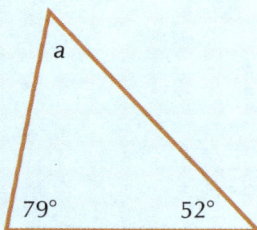

The angles of a triangle add up to 180 degrees. ⟶ Another important fact.

79° + 52° = 131°

a = 180° − 131° = 49°

EXERCISE 7B

The diagrams in this exercise are not drawn accurately.

1 Calculate the lettered angles.　**Hint:** the angles on a straight line add up to 180 degrees.

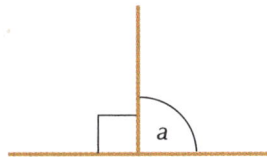　　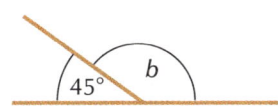　　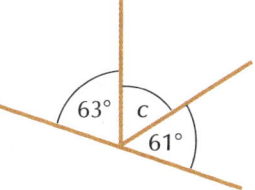

a = _____　　　　b = _____　　　　c = _____

2 Calculate the lettered angles.　**Hint:** the angles around a point add up to 360 degrees.

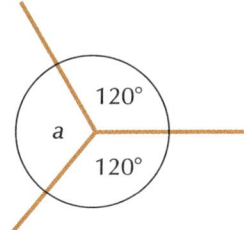　　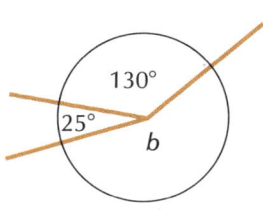　　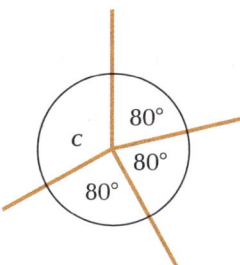

a = _____　　　　b = _____　　　　c = _____

3

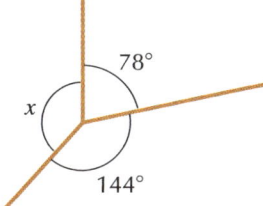　　　　Not drawn accurately

i What types of angle are 78° and 144°?

Answer 78° is _____

144° is _____　　　　　　　　　　　(2)

ii Work out the value of x.

Answer x = _____ degrees　　　　　　　　　(2)

b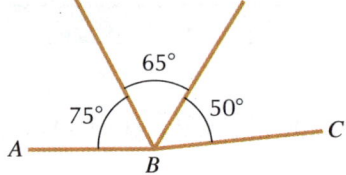

Not drawn accurately

Jasmine says that if this diagram was drawn accurately then *ABC* would be a straight line.

Is she right?

You must explain your answer.

_____

_____ (2)

(Total 6 marks)

AQA, June 2006, Paper 2 Foundation, Question 13

4 The angles of any triangle add up to 180°.

Use this fact to calculate the unknown angles in these diagrams.

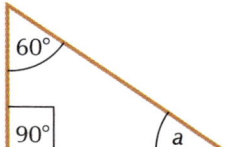

$a = $ _____

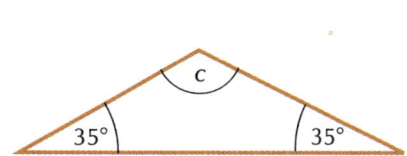

$c = $ _____

$b = $ _____

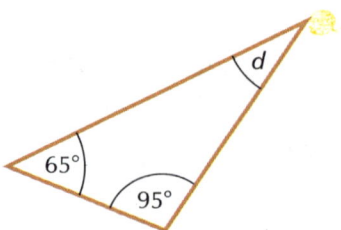

$d = $ _____

Extension

1 **a** Measure the four angles in this quadrilateral and add them up.

Total: _____

b Juno says that the four angles of a quadrilateral add up to 360 degrees.

Do you think this is true?

_____

c The four angles of a quadrilateral do add up to 360°.
On a new piece of paper draw a quadrilateral of your own.
Add up the angles and see if your answer is close to 360°.

Compare your answer with your neighbour's.

2 These three triangles are all congruent. That means they are the same size and shape.

Can you see from this diagram why the three angles of the triangle must add up to 180°? Explain how you know.

_____

_____

_____

_____

Looking at angles

In this section you will learn how to:
● find angles in real situations and in patterns.

Key words
angle
tessellation

EXERCISE 7C

FM 1 The Health and Safety Executive says that the safe angle for a ladder is at 75° to the ground.

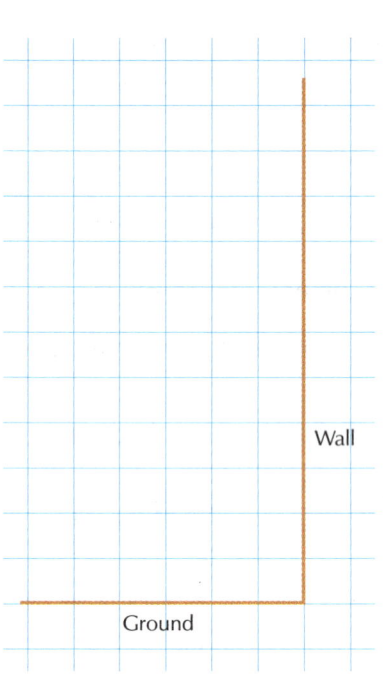

Wall

Ground

a Draw a diagram on the grid above to show a ladder leaning against a wall, measuring the 75° angle accurately.

b What is the angle between the ladder and the wall? _____

c The 'rule of 4' says that for a safe ladder, the height up the wall should be no more than four times the distance of the foot from the wall.

Check if this is true for your drawing.

2 You often see triangular frameworks in cranes and bridges.

a How many equilateral triangles are in this framework?

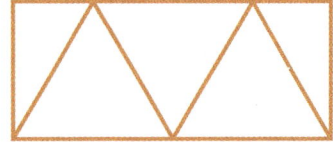

_____

b Write in the sizes of all the angles within the framework.

c Why do you think triangles are used in frameworks instead of squares or some other shape?

_____

_____

3 This is a tessellation of regular hexagons.

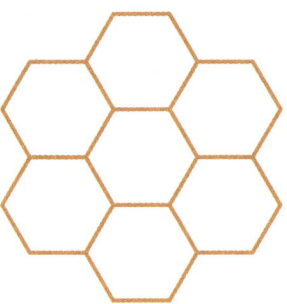

How does this show that every angle of a regular hexagon is 120°?

Hint: look at the angles around one point.

_____

_____

Extension

You could work with a partner on this exercise.

  Here is a regular octagon.

AU

PS

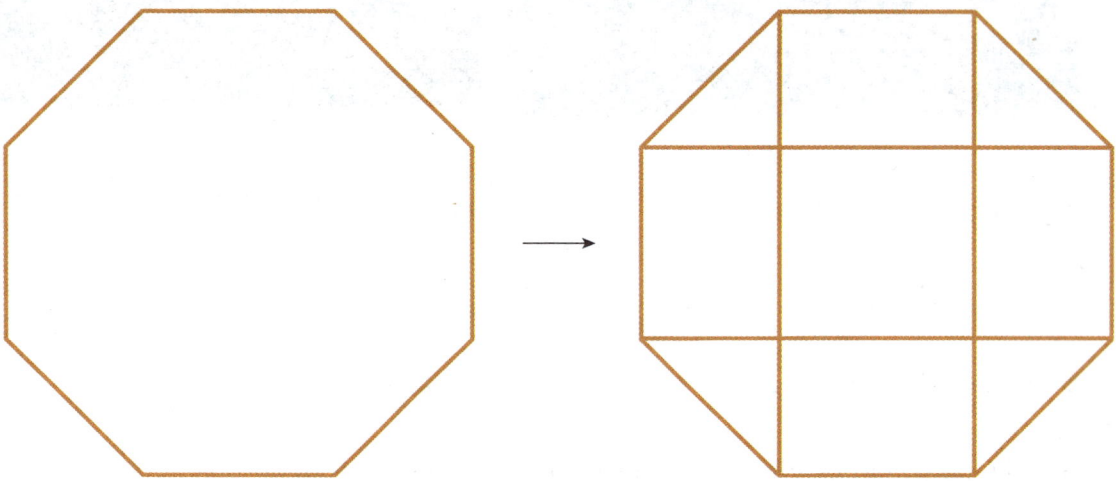

How big is each angle of the regular octagon? _____

Explain how you know. _____

_____

_____

_____

_____

_____

PS 2

a What angles surround each point in this tessellation? Show that they add up to 360°.

_____

_____

b Extend the tessellation. Colour or shade the equilateral triangles.

You could work with a partner on this activity.

This is a 30-60-90 triangle. The angles are 30°, 60° and 90°.

In this activity you will see what other shapes you can make with 30-60-90 triangles.

You will need to make drawings of all the shapes you make.

30°

90°   60°

Task A

1   Cut four 30-60-90 triangles out of card.

    They should all be the same size.

    If you are working with a partner you can share the work.

2   Put **two** of your triangles together to make an equilateral triangle.

    Make a sketch of your solution and for this new shape write in the size of each angle.

3   Put **two** of your triangles together to make an isosceles triangle. There are two ways to do this.

    Sketch both ways and write in the size of each angle of your isosceles triangles.

4   Put **two** of your triangles together to make a **parallelogram**. How many different ways are there to do this?

    Sketch each one you find and write in the size of each angle.

Task B

1   What **four-sided shapes** can you make with **three** of your triangles?
    Sketch all the possibilities and label the angles.

2   Explore the possibility of making **four-sided** shapes with **four** of your triangles.

    Keep a record of each shape and find the angles each time.

3   Using **four triangles**, experiment to see whether you can make **pentagons** or **hexagons**.
    Record your results and calculate the angles each time.

4   Add up the angles of the four-sided shapes you found in questions 1 and 2. What do you notice? Is there a similar result for pentagons or hexagons?

8 Geometry: Circles

8.1 Drawing circles

In this section you will learn how to:

- draw circles and use vocabulary connected with them.

Key words

arc	sector
chord	tangent
diameter	
perpendicular	
radius	

WORKED EXAMPLE

Draw a tangent to this circle at the point P.

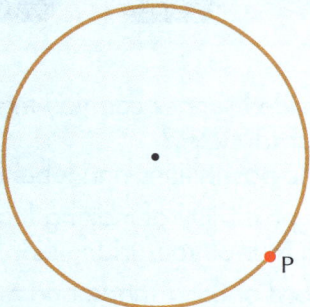

Solution

Start by drawing a **diameter** from P. This will go through the centre of the circle.

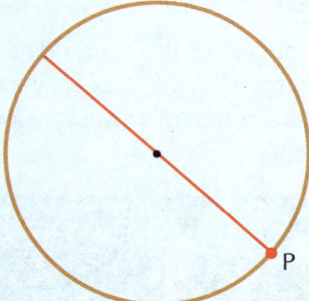

The tangent is **perpendicular** to the diameter. ⟶ Perpendicular means 'at a right angle'.

FM Functional Maths   **AU** (AO2) Assessing Understanding   **PS** (AO3) Problem Solving

Use a protractor to measure 90 degrees.   Then draw in the tangent.

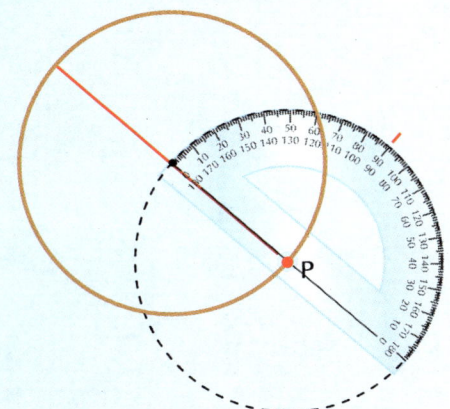   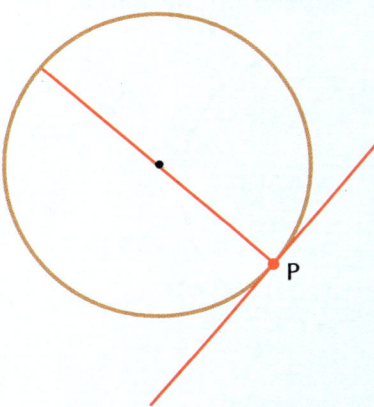

EXERCISE 8A

1 **a** Measure the radius of each of these circles.

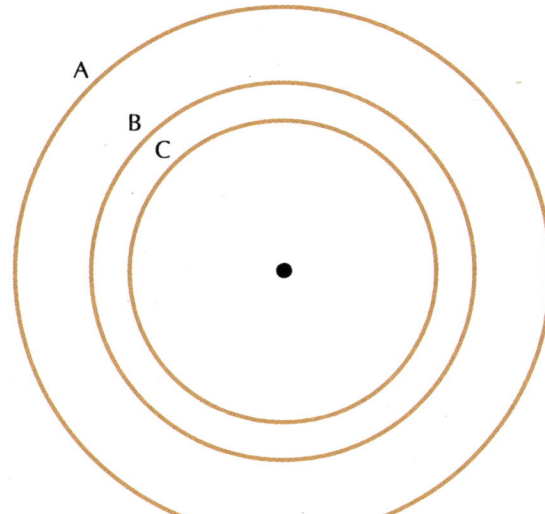

radius of A = _____

radius of B = _____

radius of C = _____

b Draw a circle of radius 3 cm on the diagram and label it D.

c What are the diameters of the four circles?

diameter of A = _____   diameter of B = _____

diameter of C = _____   diameter of D = _____

2 In each diagram, *O* is the centre of the circle.

a Draw a diameter on this circle. (1)

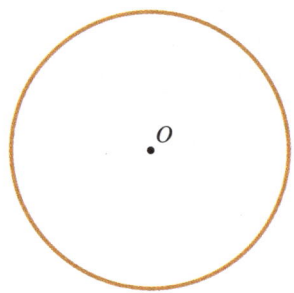

b Draw a tangent to this circle at *T*. (1)

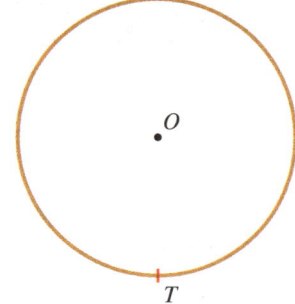

c A chord *PQ* has been drawn on the circle below.

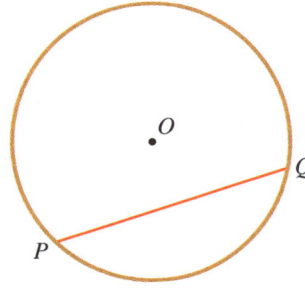

    i Mark the midpoint of *PQ* and label it *M*. (1)

    ii Join *OM*.
    What do you notice about the angle between
    OM and *PQ*?

_____ (1)

(Total 4 marks)

AQA, June 2008, Module 5, Paper 2 Foundation, Question 4

3 Fill in the gaps in these sentences.
Choose from these words:

centre, radius, tangent, diameter, chord, radius

a AC is a _____ **b** OB is a _____

c XY is a _____ **d** OA is a _____

e BC is a _____ **f** O is the _____

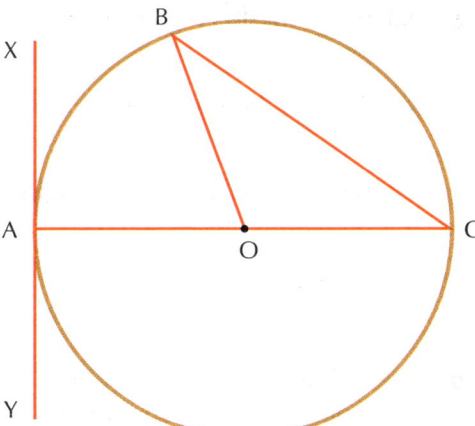

4 *O* is the centre of the circle.

A and *B* are two points on the circumference.

a Measure and write down the radius of the circle.

Answer _____ cm   (1)

b Measure and write down the size of the angle *AOB*.

Answer _____ degrees (1)

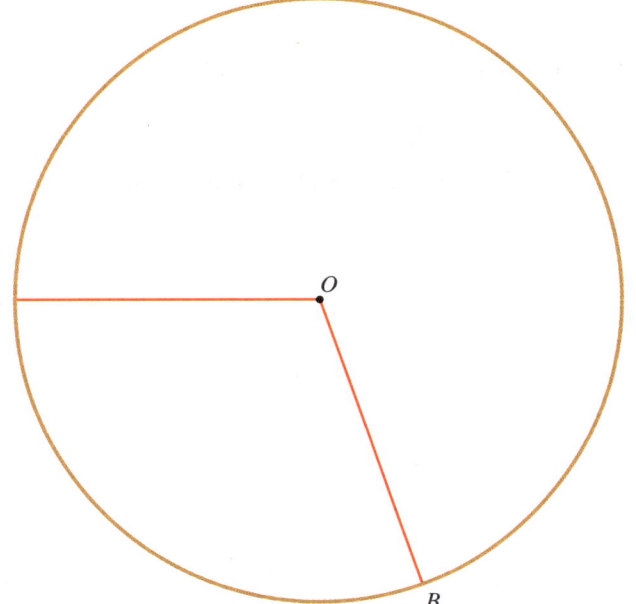

c Draw the line of symmetry of the sector *AOB*. (1)

d Draw the tangent to the circle at *A*. (1)

e Draw the chord *AB*. (1)

(Total 5 marks)

AQA, November 2008, Paper 2 Foundation, Question 4

5 Each side of this square is 6 cm.

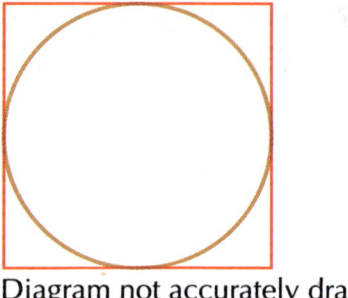

    a What is the diameter of the circle? _____

    b What is the radius of the circle? _____

Diagram not accurately drawn

6 A semicircle is **half** of a circle. On a separate sheet of paper, draw a semicircle with a diameter 8 cm long.

Extension

PS 1 This star is made by drawing four quarter circles.

See if you can draw one.

Use the dots on this grid to help you.

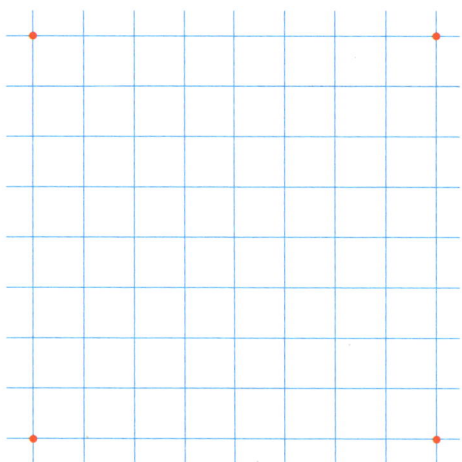

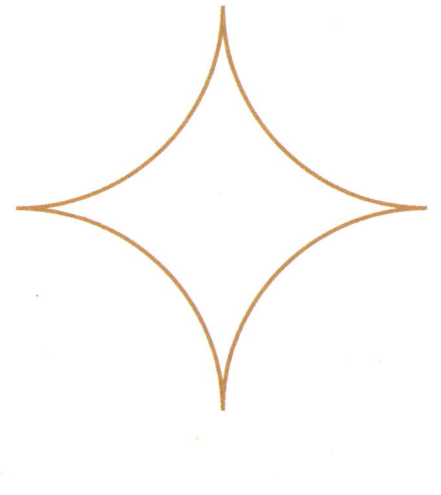

8.2 ## The circumference of a circle

In this section you will learn how to:
- use the connection between the diameter and the circumference of a circle.

Key words
circle
diameter
radius

WORKED EXAMPLE

A bicycle wheel has a diameter of 70 cm. How could we measure the circumference?
What does the circumference tell you?

Solution

The circumference is the distance all round the rim. Measuring this is difficult.
You can find it by wheeling the bike in a straight line.

Start with the valve at the bottom and stop when it
gets to the bottom again. Mark the start and finish
points on the ground.

circumference

Measure the distance between the two points.

You will find that the circumference is about 220 cm for a wheel that size.

It tells you how far the bike moves forward when the wheel goes round once.

EXERCISE 8B

You can use this graph to find the circumference of a circle,
approximately.

1 Use the graph to find the circumference of a circle
with a **diameter** of:

a 10 cm _____ **b** 5 mm _____ **c** 8 m _____

d 10 km _____ **e** 6 mm _____ **f** 7 m _____

2 Use the graph to find the circumference of a circle with
a **radius** of:

a 3 cm _____ **b** 4.5 m _____ **c** 2 cm _____

Hint: find the diameter first.

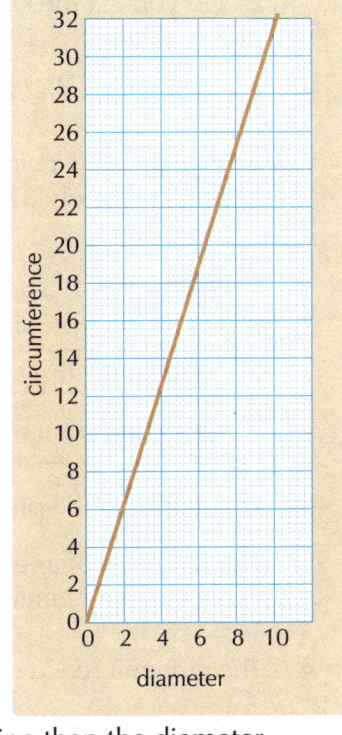

AU 3 Sometimes it is easier to measure the circumference of a drainpipe than the diameter.

The circumference of a pipe is 20 cm.

Hint: use the graph.

a What is the diameter? _____

b On a separate sheet of paper, draw a circle with a circumference of 20 cm.

c Why might it be hard to measure the diameter of a drainpipe? _____

_____

4 A park has a circular flower bed with a circumference of 14 m.

What is the diameter? _____

Extension

1 Over 2000 years ago, Archimedes proved that the circumference of any circle is approximately $3\frac{1}{7} \times$ the diameter.

a Use a calculator to find the circumference of a circle with a diameter of 10 cm by doing the multiplication $3\frac{1}{7} \times 10$. _____

b Do you get the same answer from the graph on the previous page? _____

AU 2 A circle with a diameter of 10 metres has a circumference of approximately 31.4 metres.

a Explain why a circle with a diameter of 20 metres must have a circumference of approximately 62.8 metres.

_____

b Complete this table:

Diameter (metres)	5	10	20	40
Circumference (metres)		31.4		

3 The graph on the previous page also works for spheres.

The diameter of a tennis ball is 10 cm.
What is the circumference of a tennis ball? _____

4 The circumference of the Earth is 25 thousand miles. This is the length of the equator.

FM AU a New Zealand is almost on the opposite side of the Earth to the UK. If you could travel there by the shortest route, how far would you go? _____

b What is the diameter of the Earth? _____

8.3 Circles in action

In this section you will learn how to:
- handle circles in realistic situations.

Key words
circumference
diameter
radius

EXERCISE 8C

FM 1 Measure the diameters of some circular coins.

a Record your results.

Value	1p	2p	5p	10p	£1	£2
Diameter						

b Is it true that larger coins have a bigger value? Give a reason for your answer.

_____

2 This is a simple version of the flower of life, a symbol that is found in many cultures.

Draw it here.

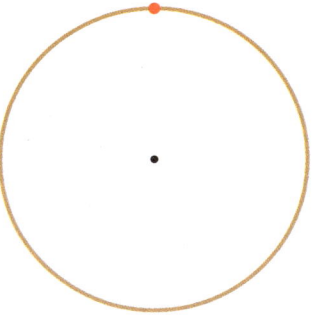

Hint: start with an arc using the point marked as the centre and the same radius as the circle.

 Drink cans could be packed together in two different ways.

These diagrams show an overhead view.

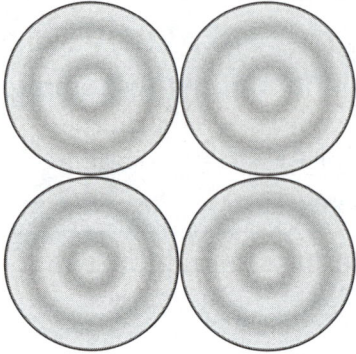

Square packing

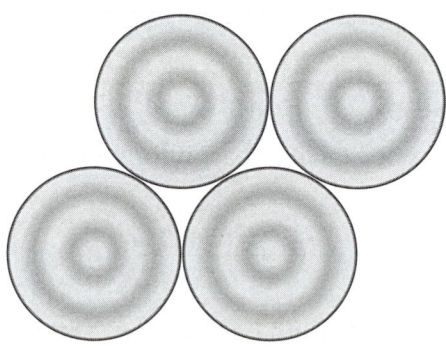

Hexagonal packing

a Which type is commonly used in packs of drinks cans?

_____

b On a separate piece of paper, draw around a coin to illustrate square packing of six cans arranged in a rectangle.

c Now draw round a coin to illustrate hexagonal packing of seven cans arranged in a hexagon.

d Which type has the least empty space between the cans?

_____

Extension

FM 1 Here is a can of baked beans. You have been asked to design a label for the can.

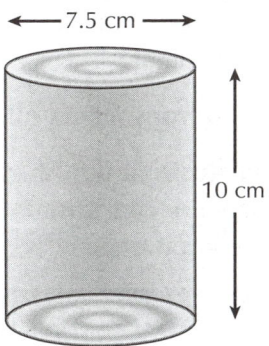

7.5 cm

10 cm

a What is the circumference of the can? _____

> **Hint:** use the graph in exercise 8B.

b The label will be a rectangle. What size will it be?

_____

The strength of the signal from a television transmitter depends on how far you are from the transmitter.

There are two transmitters 60 km apart.

Produce an illustration to show the distances from each transmitter to places in the surrounding area.

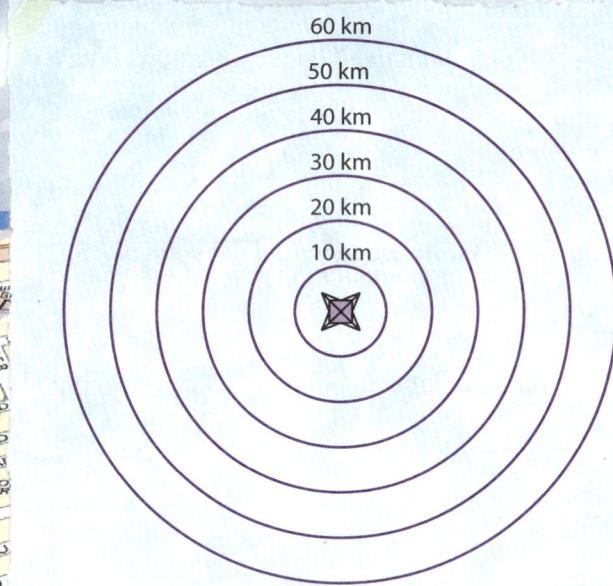

Task A

1 Mark two transmitters, A and B, 60 km apart, on a separate sheet of plain paper. Use a scale of 1 cm to 10 km.

2 Draw two sets of concentric circles, like the circle in the diagram above. Remember to use a scale of 1 cm to 1 km.

   One set will be centred on transmitter A and the other on transmitter B.

3 It will be useful to see which of the two transmitters is nearer to any particular place. Do it by working through these questions.

   a There is one point 30 km from both A and B. Mark it with a cross.

   b There are two points 40 km from A and from B. Mark them both with a cross.

   c Mark with crosses points which are:
      - 50 km from A and B
      - 60 km from A and B
      - 70 km from A and B

4 Join your crosses with a line.

   a What can you say about the line you have drawn?

   b Where are the places which are nearer to transmitter A than they are to transmitter B?

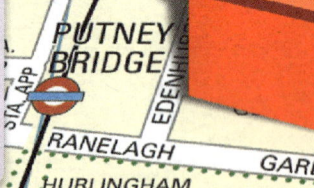

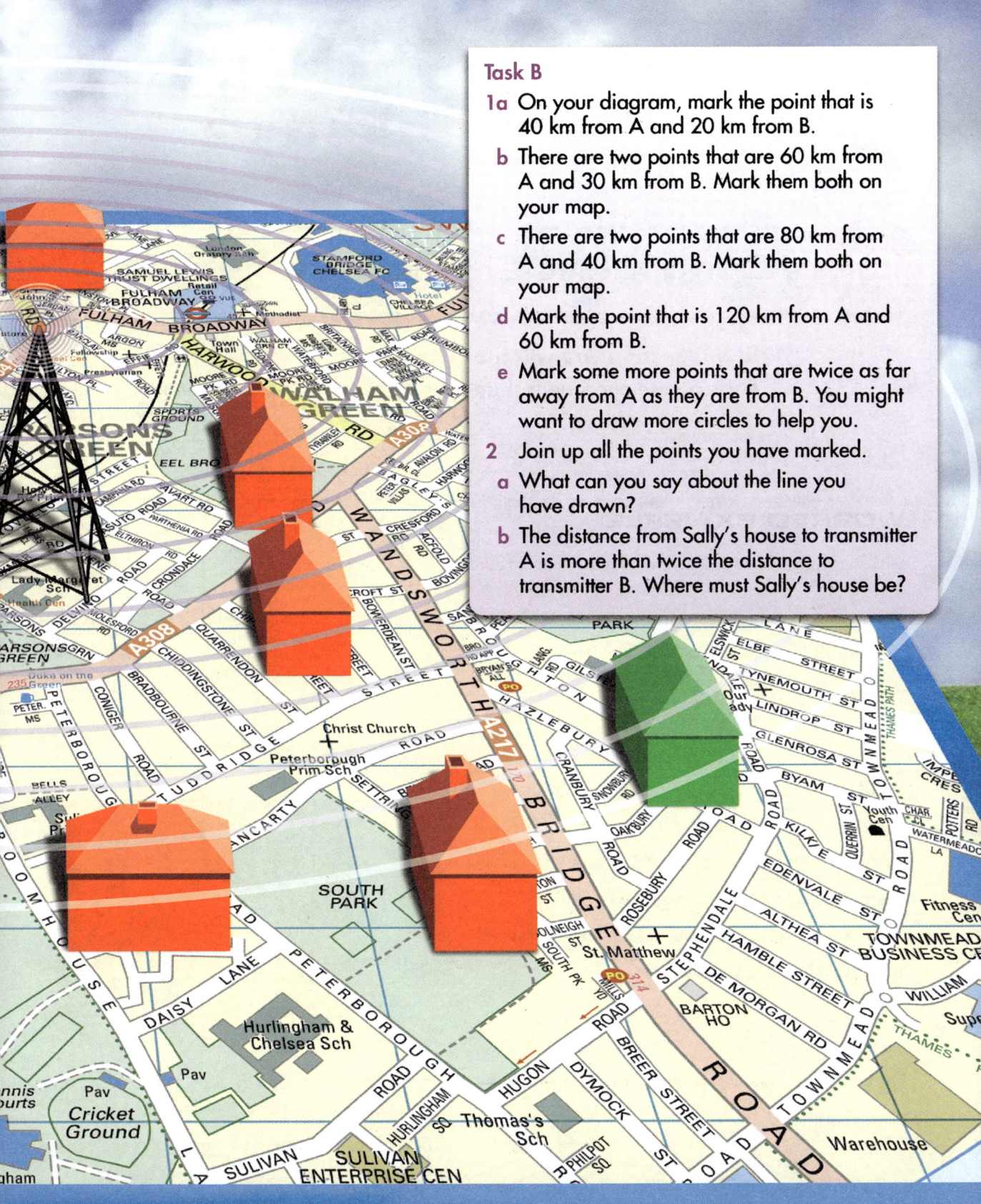

Task B

1a On your diagram, mark the point that is 40 km from A and 20 km from B.

b There are two points that are 60 km from A and 30 km from B. Mark them both on your map.

c There are two points that are 80 km from A and 40 km from B. Mark them both on your map.

d Mark the point that is 120 km from A and 60 km from B.

e Mark some more points that are twice as far away from A as they are from B. You might want to draw more circles to help you.

2 Join up all the points you have marked.

a What can you say about the line you have drawn?

b The distance from Sally's house to transmitter A is more than twice the distance to transmitter B. Where must Sally's house be?

Geometry: Transformations

9.1 Congruent shapes

In this section you will learn how to:
- recognise congruent shapes.

Key words
congruent
enlargement
reflection

WORKED EXAMPLE

Which of the shapes in this rectangle are congruent?

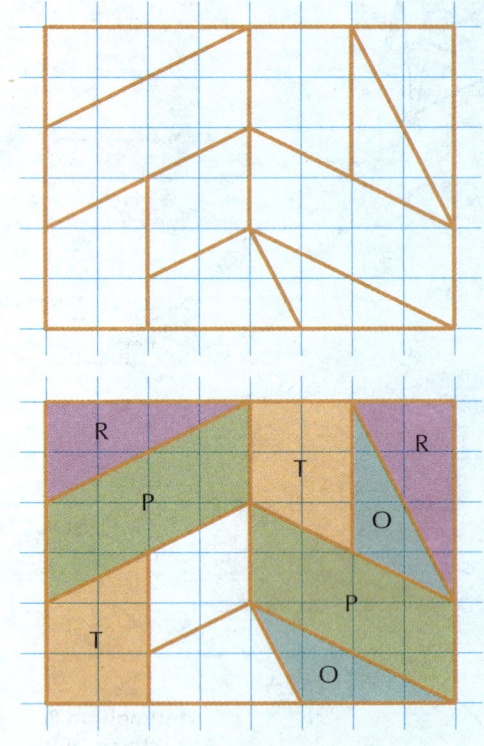

Solution

There are two congruent right-angled triangles. (R)

There are two congruent parallelograms. (P)

There are two congruent trapeziums. (T)

There are two other triangles which are congruent. (O)

They are labelled in this diagram.

The remaining two shapes are not congruent. ⟶ You can check these by copying them on paper and cutting them out.

Sometimes you need to turn one over to match them up.

FM Functional Maths   **AU** (AO2) Assessing Understanding   **PS** (AO3) Problem Solving

EXERCISE 9A

1 Two shapes are **congruent** if they are identical. One could be placed exactly on top of the other.

A  B 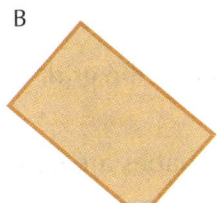 C  D

Three of these rectangles are congruent. Which is the odd one out? Why?

The odd one out is _____ because _____

2 a Match the pairs of congruent triangles.

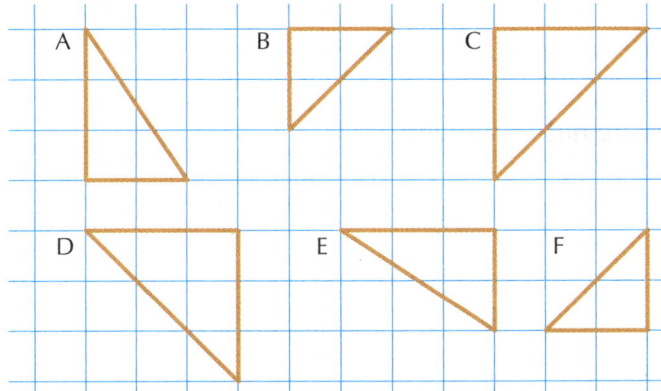

_____ and _____

_____ and _____

_____ and _____

b Which triangles are isosceles? _____

3 Answer this question with a partner.

These shapes are called **pentominoes**. They all contain five squares.

a Match up the **congruent** pentominoes.

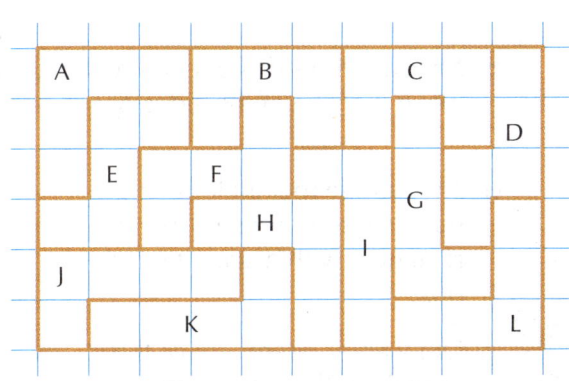

_____

_____

_____

> **Hint:** if you are not sure, copy the shapes and cut them out.

b Which is the odd one out? Why? _____

> **Hint:** colour congruent pentominoes in the same way.

AU **4** **a** What is the name of this shape? _____

b Draw in the reflection on the grid.

c Explain why the two shapes are congruent.

_____

_____

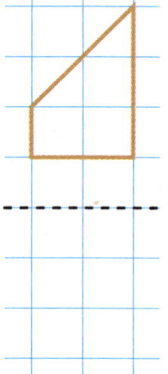

Extension

1 **a** Explain why these rectangles are **not** congruent.

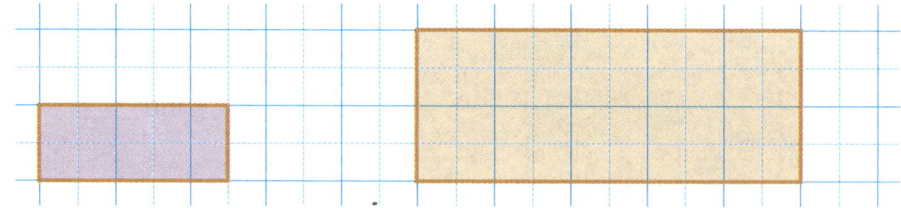

_____

_____

b Measure the sides of each rectangle and check that the sides of the orange rectangle are twice the size of the sides of the purple rectangle.

c The orange rectangle is an enlargement of the purple rectangle with a scale factor of 2.

Draw an enlargement of this triangle with a scale factor of 2.

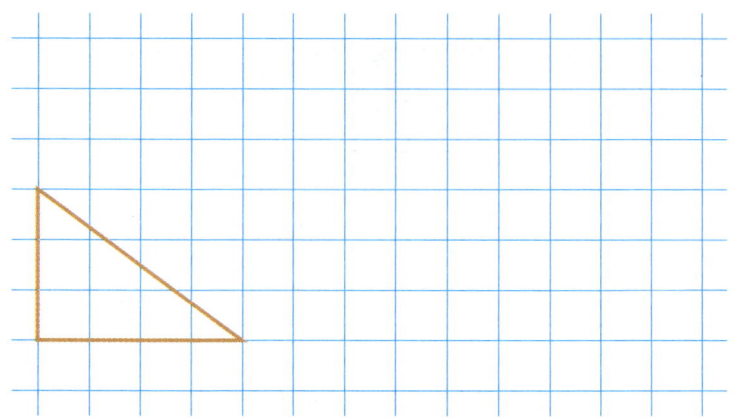

2 Draw your own example of an enlargement.

9.2 Putting shapes together

In this section you will learn how to:
- fit shapes together in tessellations.

Key words
congruent
tessellation

WORKED EXAMPLE

Show how to make a tessellation from this shape.

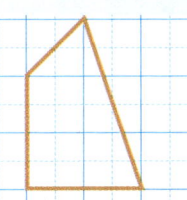

Solution

First rotate the original tile by half a turn.

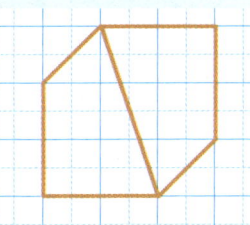

Now repeat these two tiles.

We can repeat this for as long as we wish.

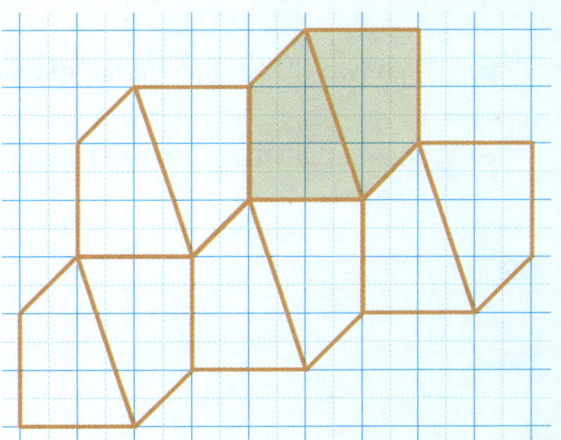

EXERCISE 9B

1 Here is one way to divide a square into four congruent parts.

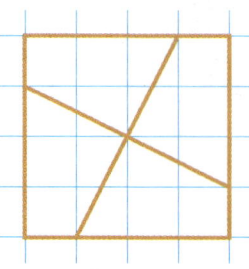

Show two other ways to divide a square into four congruent parts.

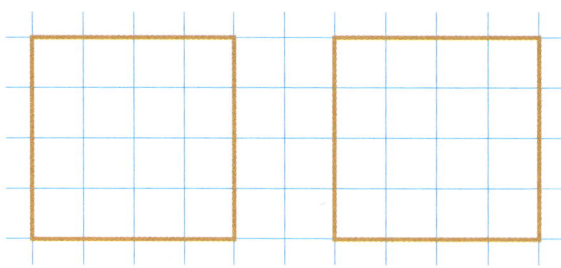

2 Here are six shapes made from centimetre squares.

A   B   C

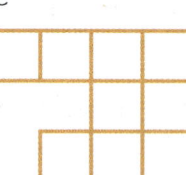

D   E   F

a Which two shapes fit together to make a rectangle? _____ and _____ (1)

b Which two shapes fit together to make a square? _____ and _____ (1)

c Work out the area of shape D. State the units of your answer.

_____ _____ (2)

(Total 4 marks)

AQA, May 2009, Paper 1 Foundation, Question 4

AU 3 A trapezium is being used to make a regular pattern.

A pattern like this is called a **tessellation**.

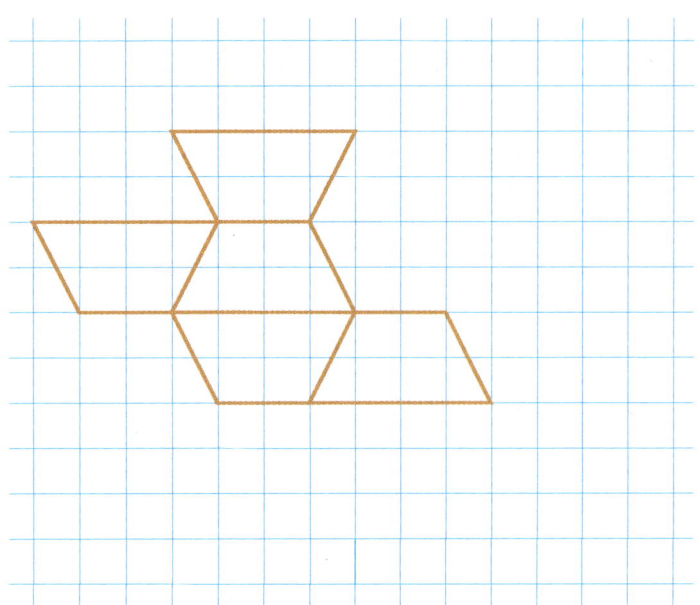

a Show how the pattern could continue.

b Bathroom, kitchen and floor tiles are usually square or rectangular.

Could trapezium-shaped tiles be used?

_____

c Why do you think trapezium-shaped tiles are not used? _____

_____

Extension

PS 1 This is the start of a tessellation. The four coloured tiles are all congruent.

Continue the pattern and colour it in.

Hint: use the lines to help you.

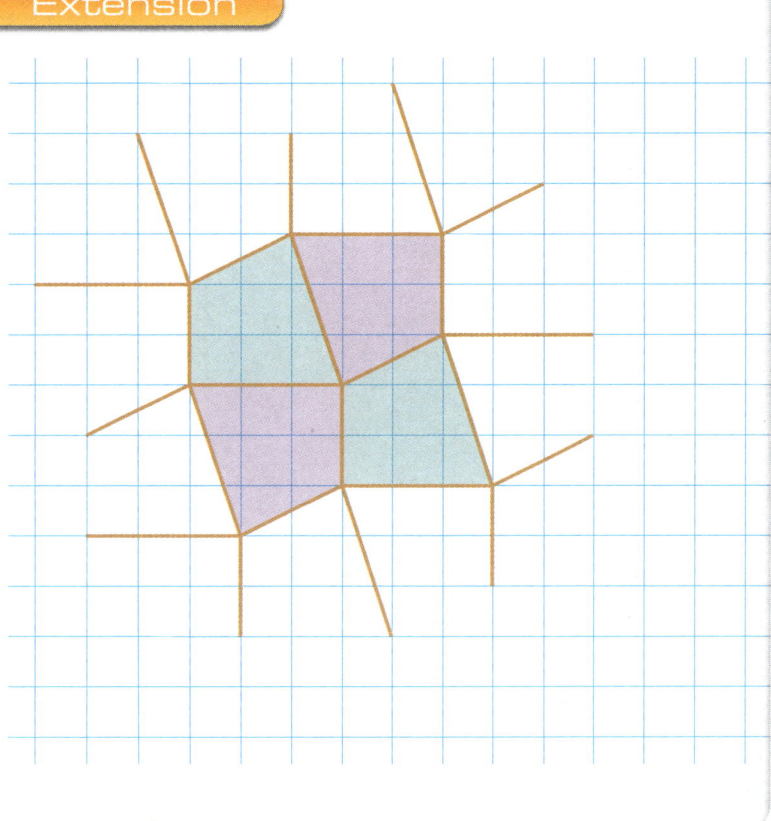

2 Here are some shapes and their scale 2 enlargements.

In each case, show how four of the original shapes will fit in the enlargement. The first one has been done for you.

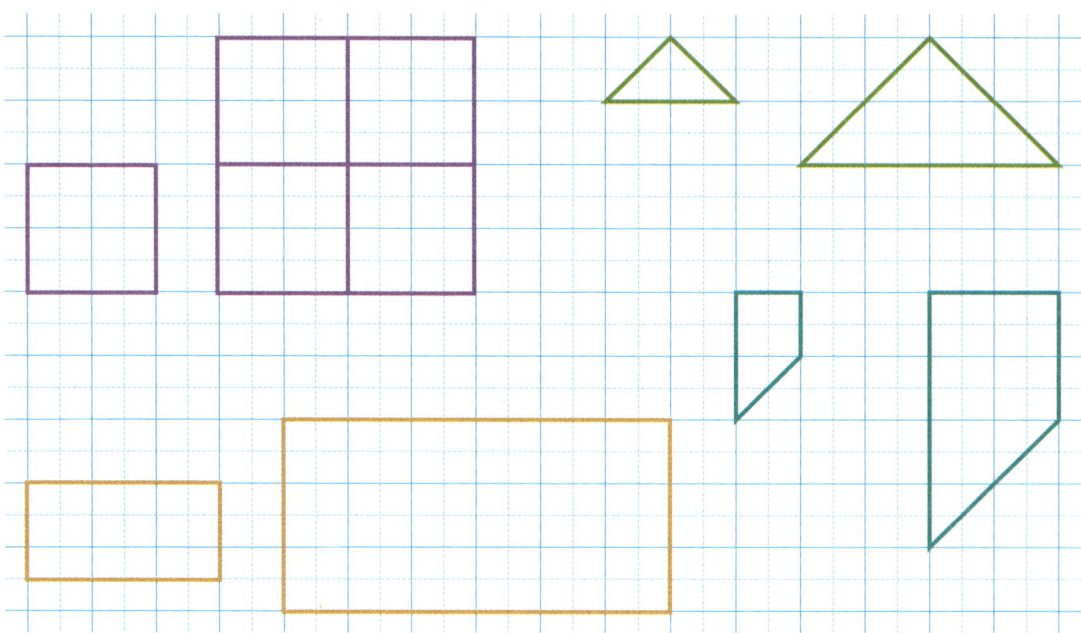

3 Show how to fit four of the shapes on the left into the square on the right.

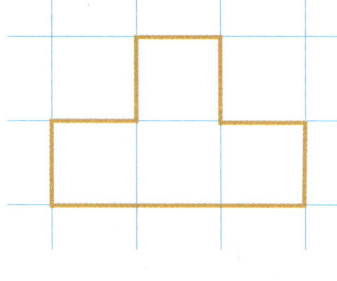 

9.3 Using patterns from simple shapes

In this section you will learn how to:
● identify patterns in everyday situations.

Key words
congruent
tessellation

EXERCISE 9C

 1 A farmer has an orchard with 12 trees.

She wants to leave it to her four daughters so that each one has the same shaped piece of land and the same number of trees.

Show on the diagram how she can do this.

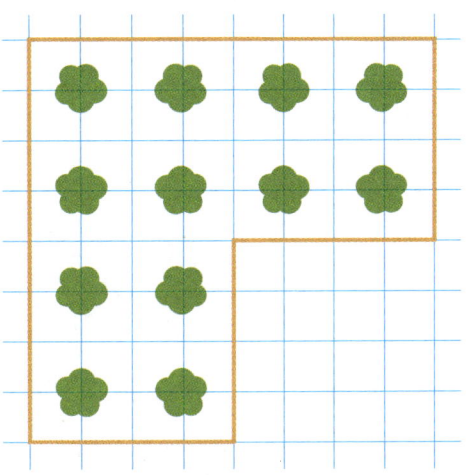

2 This is called a herringbone pattern. It can be found in brickwork or in cloth.

This is the start of a herringbone pattern.

Continue the pattern.

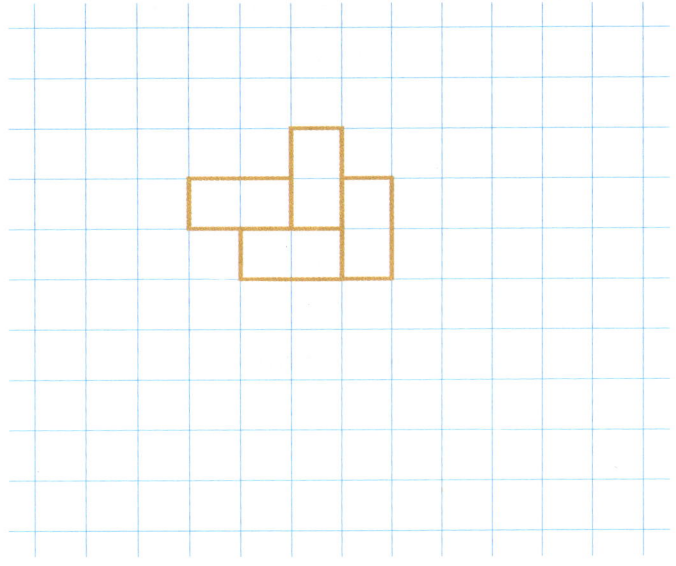

FM 3 A patio is to be covered with a mixture of square and octagonal tiles.

It will start like this.

Help the owner of the house to get an idea of what it will look like when finished by continuing the pattern.

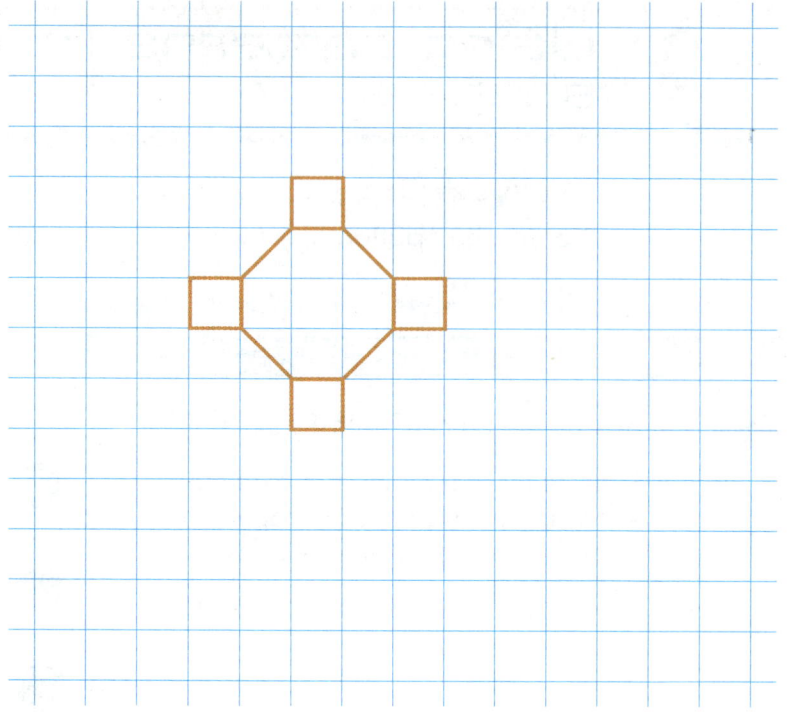

4 Square tiles are often designed to show a pattern when they are put together.

Here is a simple design on one tile.

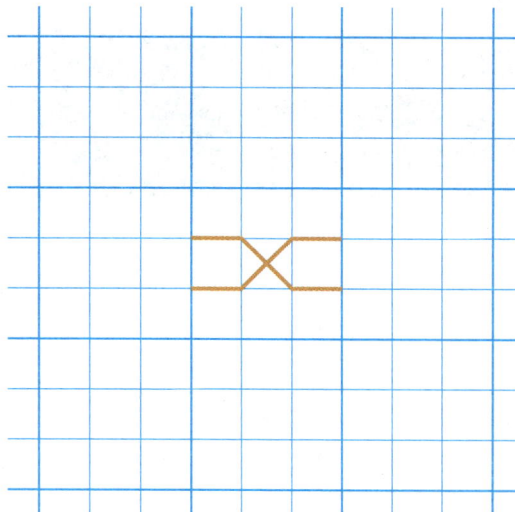

Copy the pattern onto the surrounding tiles.

Colour each tile in an identical way.

Extension

1 Make up your own tile design
and repeat it in each large square.

Colour each tile in an identical way.

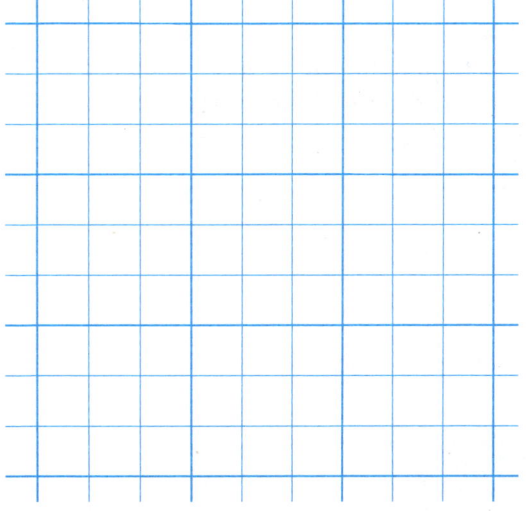

2 This design was found on a textile made in Japan in the 18th century.

a What symmetry does this pattern have? _____

b This design is repeated across the whole of the Japanese textile.
Show how this could be done.

checklist

☐ I can recognise congruent shapes.

☐ I can fit shapes together in tessellations.

☐ I can identify patterns in everyday situations.

Problem Solving
Pentomino puzzles

You could work with a partner on this task.

A shape that is made by putting five squares together edge-to-edge is called a pentomino. You will be familiar with these if you have played the computer game Tetris.

Task A

This rectangle is made of pentominoes.

In the rectangle there are six different pentominoes.
E and F are **not** different because they are **congruent**.
If you turn E over you can put it on top of F.

1 Match up the congruent pentominoes in the rectangle.
   If you are not sure cut some out of paper to check.

2 Which are the two which are not congruent to any others?

Task B

1 It is possible to make 12 different pentominoes. Can you find them all?

2 Which pentominoes look the **same** when you turn them over?

Here is a 5 by 6 rectangle. It has been filled with six copies of a single pentomino.

3 What other pentominoes can be used to fill the 5 by 6 rectangle **on their own**? Draw your own grids like the one below and make drawings to show how this can be done.

Task C

Here are two pentominoes.

1 Show how pentominoes of these two shapes **together** can fill a 5 by 6 rectangle.

2 What other **pairs** of pentominoes can be used together to fill a 5 by 6 rectangle? Make drawings to show how.

3 Can you fill a 5 by 6 rectangle with six **different** pentominoes?

4 A manufacturer wants to make a pentomino puzzle.

It will have the 12 different pentominoes made out of plastic.

He wants to fit them into a 6 by 10 rectangular box (like the rectangle in Task A).

The trouble is he cannot find a way to fit them in! Can you help?

Hint: this is not easy! You may want to cut out copies of the 12 pentominoes to experiment with.

10.1 Drawing triangles

In this section you will learn how to:
- draw triangles accurately using given measurements.

Key words

angle
triangle

WORKED EXAMPLE

Draw this triangle accurately.

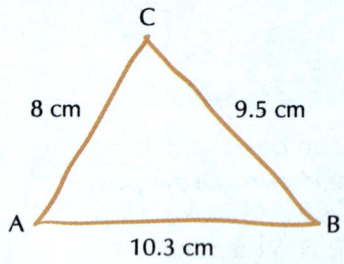

Solution

Start by drawing the line from A to B 10.3 cm long.

Next, open your compasses to 8 cm.

Put the point on A and draw an arc.
This is drawn to a smaller scale here.

You will need a ruler and a pair of compasses.

Now open the compasses to 9.5 cm.

Put the point on B and draw a second arc.

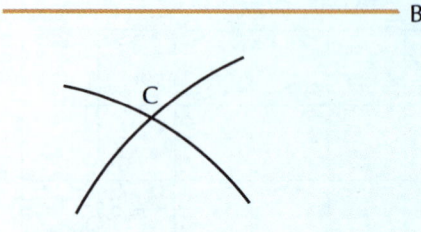

FM Functional Maths   **AU** (AO2) Assessing Understanding   **PS** (AO3) Problem Solving

Where the arcs cross is point C.

Draw in the other two sides.

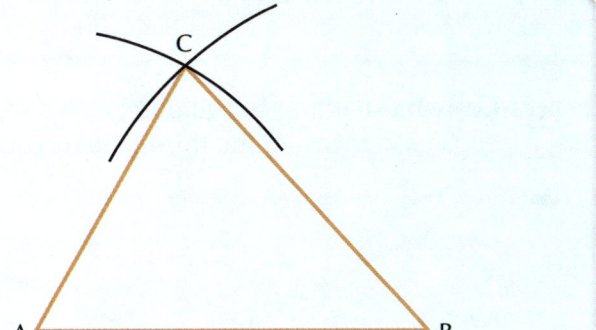

Check the sides by measuring them.

EXERCISE 10A

1 Measure the lengths of these lines in centimetres.

a _____

　　Length = _____ cm

b _____

　　Length = _____ cm

c _____

　　Length = _____ cm

2 Measure the lengths of these lines and the angle between them.
Label these on the diagram.

Extension

1 Draw these triangles accurately on a separate piece of paper. Use a ruler and a protractor. In each case measure the third side in your drawing. The triangles are not drawn accurately.

a

4 cm

80°

8 cm

b

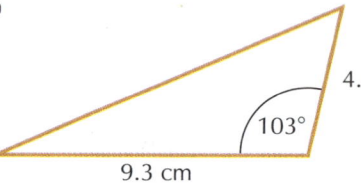

4.6 cm

103°

9.3 cm

c

10 cm

38°

8.2 cm

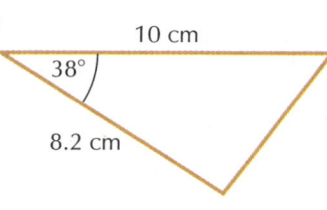

a _____ cm   **b** _____ cm   **c** _____ cm

2 Draw these triangles on a separate piece of paper. They are not drawn accurately here.

a

65°   40°

8 cm

b

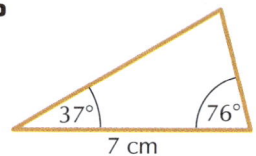

37°   76°

7 cm

c

60°   55°

6.5 cm

Hint: start by drawing the base line accurately. Then measure the angle and draw a line at each end.

d

35°   35°

11 cm

e

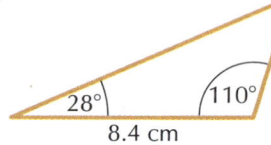

28°   110°

8.4 cm

f

60°   60°

7.5 cm

3 Look at the triangles in question 2.

a What type of triangle is **d**? _____

b Measure the other two sides on your drawing of triangle **d**. They should be the same length.

c What type of triangle is **f**? _____

d How long should the other sides of triangle **f** be? _____

e Measure the sides of triangle **f** to check your accuracy.

You will need a pair of compasses for these questions.

4 Draw these triangles accurately on a separate piece of paper.

Hint: use the method shown in the example.

a

3.5 cm   4 cm

4.5 cm

b

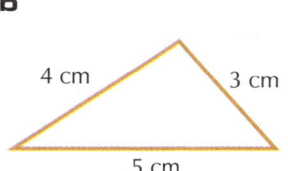

4 cm   3 cm

5 cm

5 If you have drawn triangle **b** in question 4 accurately, one of the angles should be a right angle.
Measure it to check your accuracy.

U 6 **a** Draw this triangle accurately on a separate piece of paper.

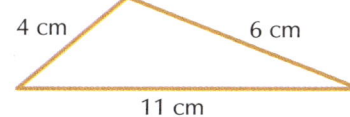

4 cm   6 cm

11 cm

b Explain what happens.

_____

_____

c Make your own sketch of an impossible triangle.

7 Here is a sketch of a triangle.

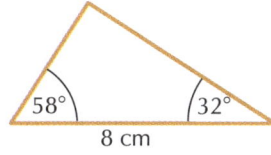

58°   32°

8 cm

In the space below, make an accurate drawing of the triangle.

(Total 3 marks)

AQA, June 2006, Paper 1 Foundation, Question 13

10.2 Bearings

In this section you will learn how to:
● use a bearing to describe a direction.

Key words
angle
bearing
clockwise
direction

WORKED EXAMPLE

The scale of a map is 1 cm to 20 km.

Freshton is 90 km from Walden at a bearing of 260 degrees.

Mark it on the map.

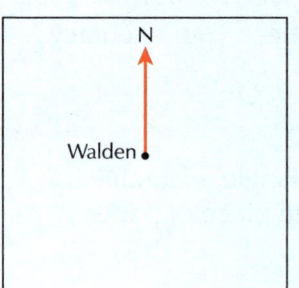

Solution

Draw a line from Walden. The bearing is the angle clockwise from north.

It is over half a complete turn because it is more that 180 degrees.

260 − 180 = 80 so we need to measure 180 + 80 degrees.

Use a protractor.

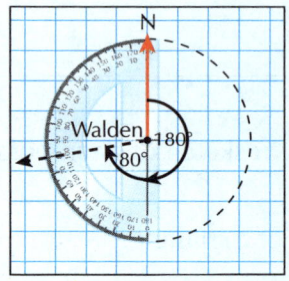

The distance is 90 km. On the map that is 90 ÷ 20 = 4.5 cm.

Measure 4.5 cm from Walden to find Freshton.

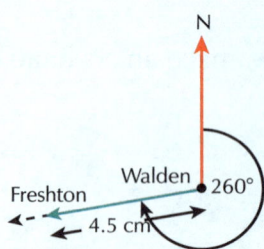

EXERCISE 10B

1 Find the bearing from Aybury of these towns.

Bearings are written with three digits so the bearing of Besley from Aybury is 050°.

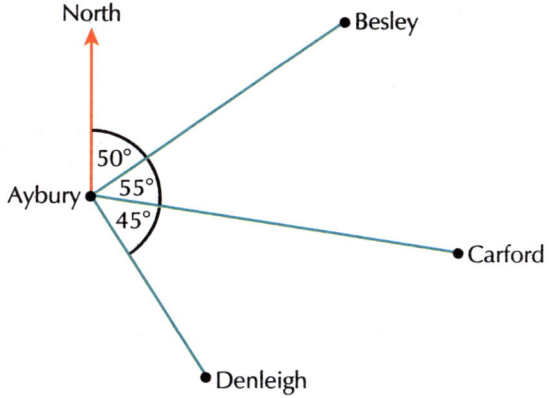

a Carford _____

b Denleigh _____

Hint: the drawings are not accurate. You cannot use a protractor.

2 Find the bearing of the other landmarks from the station.

hill: _____ °

mosque: _____ °

wood: _____ °

church: _____ °

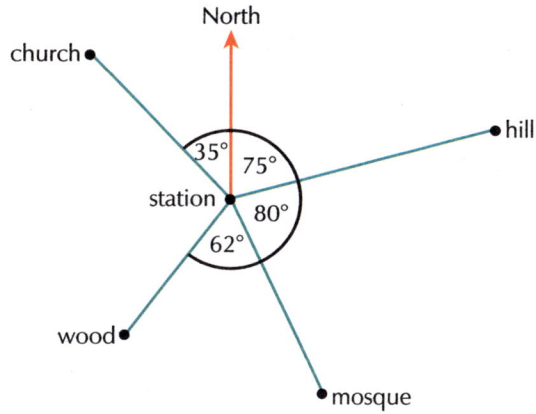

3 **a** Find the bearing of Eastley from Weston. _____ °

b Find the bearing of Weston from Eastley. _____ °

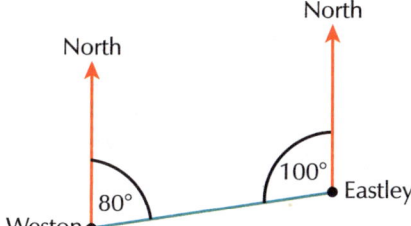

Extension

1 The scale of this map is 1 cm to 5 miles.

● Farnbury

a Measure the distance from Castorleigh to Farnbury.

North

_____ miles

b Find the bearing from Castorleigh to Farnbury.

Castorleigh ●

_____ °

c The bearing from Castorleigh to Walbridge is 110°. Draw a line in this direction on the map.

d The distance from Castorleigh to Walbridge is 40 miles. Mark Walbridge on the map.

10.3 Maps and journeys

In this section you will learn how to:
● use bearings in realistic situations.

Key words
angle
bearing
direction

EXERCISE 10C

S **1** Bearings are used by planes and ships to specify the direction to travel.

Write down the bearings of each of these compass directions.

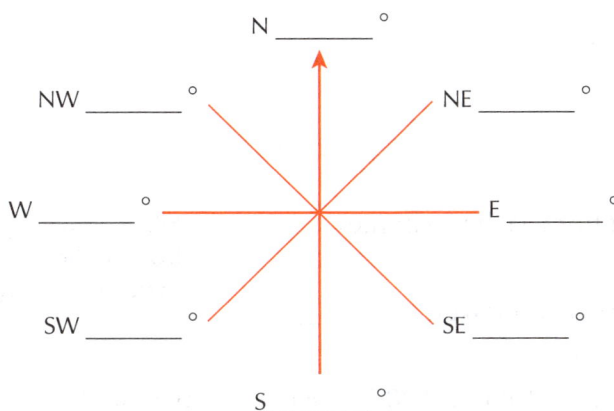

2 This map shows three airports in France and Spain.

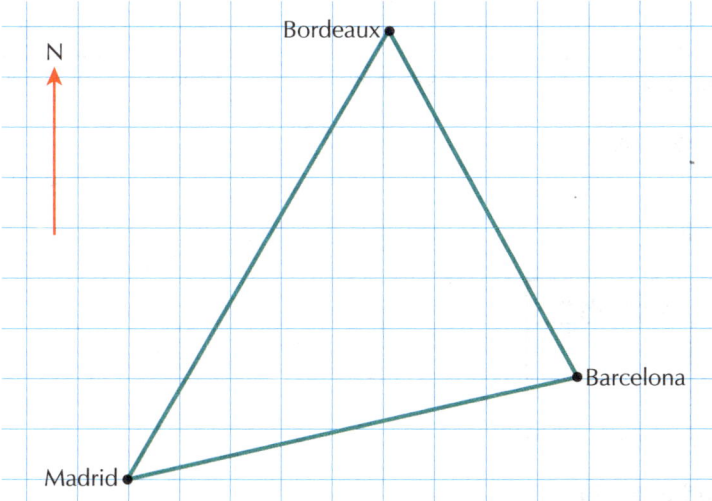

On the map, mark angles to show these bearings.

a The bearing of Bordeaux from Madrid is 030°.

b The bearing of Barcelona from Bordeaux is 152°.

c The bearing of Madrid from Barcelona is 257°.

3 This map shows two islands in the Mediterranean Sea.

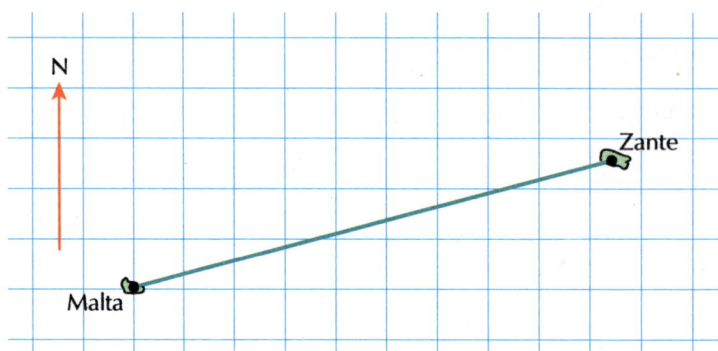

You are taking a boat from Malta to Zante.

Hint: draw a north line from Malta to help you measure the angle.

a On what bearing will you travel? _____°

b The scale of the map is 1 cm to 100 km. How far apart are the islands?

_____

c If you travel from Zante back to Malta, what is the bearing for this journey? _____°

FM 4 This chart shows two stages of an aircraft's flight from Glasgow to Manchester.

a Describe the distance and bearing for each stage of the journey.

Put your answers in this table.

	Bearing	Distance
Glasgow to turn		
turn to Manchester		

Hint: draw north lines at Glasgow and at the turn.

b Why do you think the plane did not fly directly to Manchester?

_____

_____

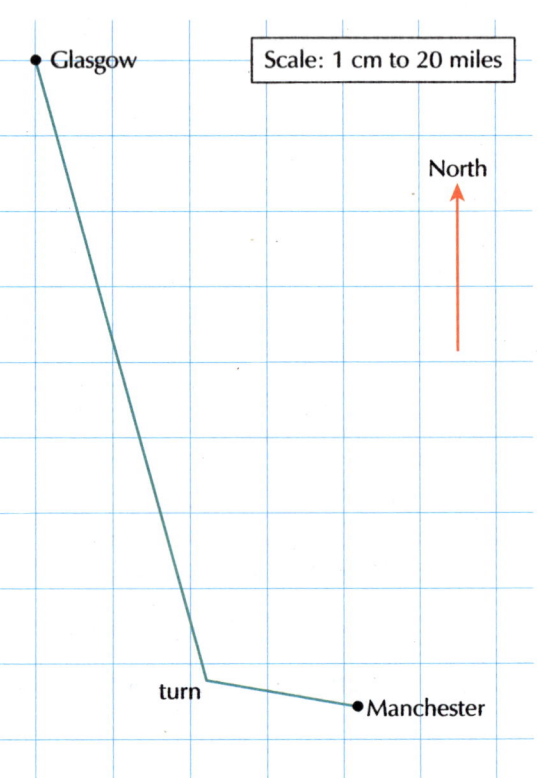

Extension

1 Here is Birmingham airport marked on a map. The scale of the map is 1 cm to 20 miles.

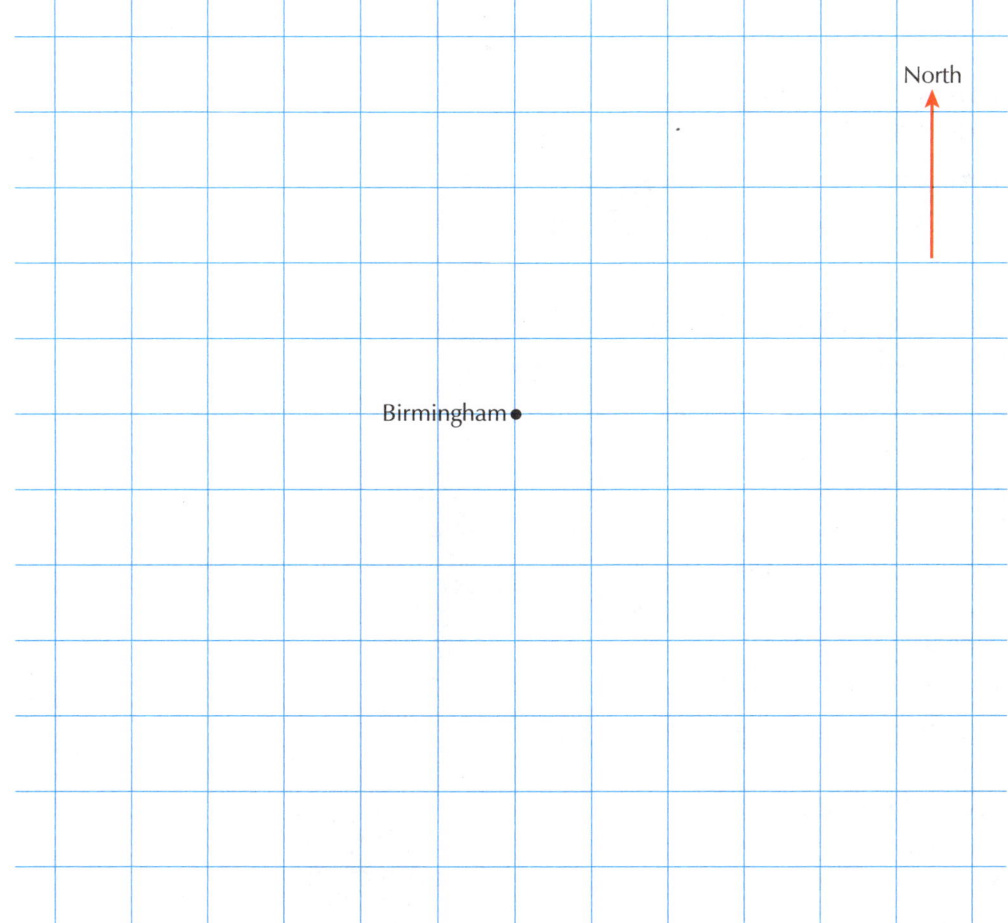

This table shows the distance and bearings of some other airports from Birmingham. Show the airports on the map.

Compare your finished map with your neighbour's.

Airport	Bearing from Birmingham	Distance from Birmingham
Gatwick	145°	116 miles
Leeds-Bradford	010°	94 miles
Southampton	168°	106 miles
Cardiff	225°	90 miles
Liverpool	327°	84 miles

checklist

☐ I can draw triangles accurately using given measurements.

☐ I can use a bearing to describe a direction.

☐ I can use bearings in realistic situations.

Problem Solving
Yacht race

This is a game for two or more people.

You will need:

- Two dice
- A small counter to mark your position (a 5p coin works well)

For the **distance** dice, move 1 km for each dot on the dice. Here are the instructions for the **direction** dice:

Throw	Bearing
1	060°
2	120°
3	180°
4	240°
5	300°
6	360°

Instructions

The aim is to be the first person to sail round the island from the start to the finish.

Yachts move from dot to dot. The dots are 1 km apart.

- To move you throw **two dice**.
- One will give the **direction** to move your yacht and the other will give the **distance**.
- **You can choose** which dice to use for the direction and which to use for the distance.
- You cannot sail on or across the island.
- You cannot finish on the same point as another yacht.
- You cannot sail off the board.
- If you cannot move, you must miss a turn. You must move if you can.
- You can win by sailing over the finishing point. You do not have to stop there.
- If you complete one race, have a second, but this time go from the finish back to the start.

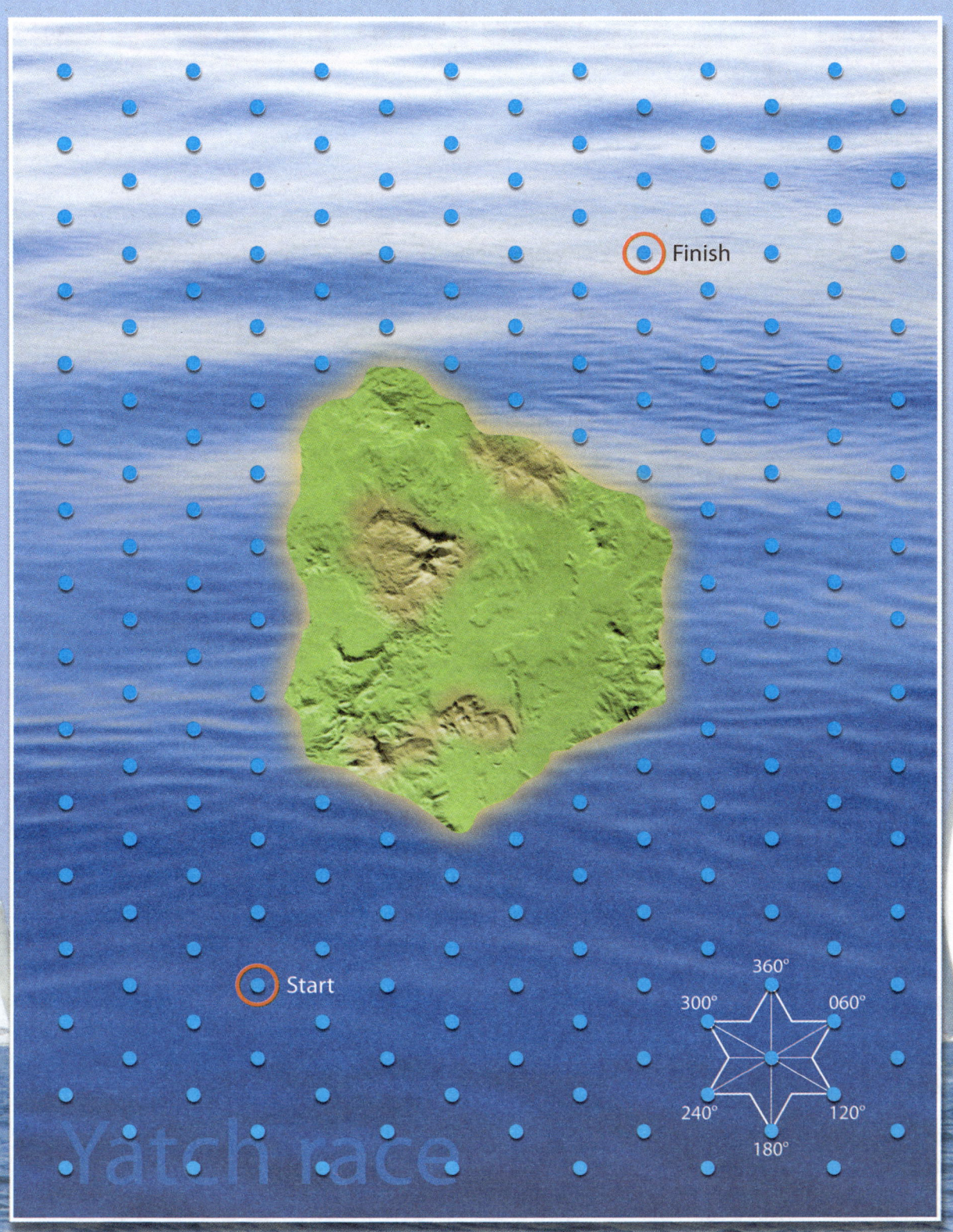

Finish

Start

360°

300°　060°

240°　120°

180°

Yatch race

11 Geometry: Volume

11.1 Shapes made from cubes

In this section you will learn how to:
- find the surface area and volume of shapes made from cubes.

Key words
centimetre cube
cube
surface area
volume

WORKED EXAMPLE

Find the volume and surface area of this shape made from centimetre cubes.

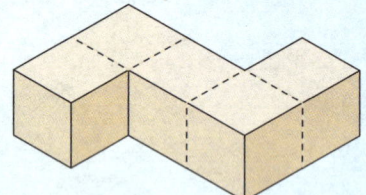

Solution

The shape is made from five centimetre cubes, so the volume is 5 cm^3.

The surface area is the total of the areas of each face.

The top has an area of 5 cm^2. ────────→ Just count the squares.

The bottom is also 5 cm^2.

Around the edge there are 12 squares, ────→ You can see 6 squares in the diagram
so the area is 12 cm^2. and another 6 are hidden.

The total surface area is 5 + 5 + 12 = 22 cm^2. ──→ Make the shape from cubes if you are
not sure about this.

EXERCISE 11A

It will help if you have some cubes when you are working through this chapter.

You can clip them together to make the different shapes.

The diagrams in this section are not drawn to scale.

1 Here is a shape made out of centimetre cubes.

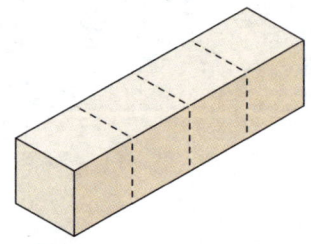

   a How long is it? _____ cm

FM Functional Maths   **AU** (AO2) Assessing Understanding   **PS** (AO3) Problem Solving

b What is the volume? _____ cm^3

c The area of the top is 4 cm^2. What is the **total** area of all **six** faces? _____ cm^2

2 Here is a different arrangement of the cubes in question 1.

a What is the volume? _____ cm^3

b How many faces does it have? _____

c What is the **total** area of **all** the faces? _____ cm^2

3 On a separate piece of paper, draw a different arrangement of four centimetre cubes and find the total surface area.

Surface area = _____ cm^2

4 Here are two different shapes made with six centimetre cubes.

A   B

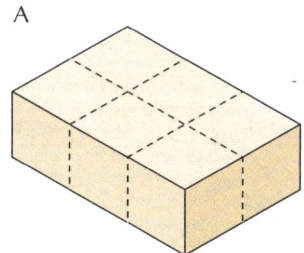
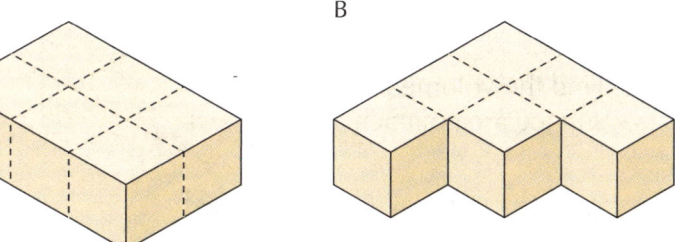

a Find the volume and surface area of shapes A and B.

A   Volume = _____ cm^3 Surface area = _____ cm^2

B   Volume = _____ cm^3 Surface area = _____ cm^2

b Find an arrangement of six cubes which has a larger surface area than shape B.

Surface area = _____ cm^2

5 A strip of paper one centimetre wide is wrapped around the outside of this shape made from centimetre cubes.

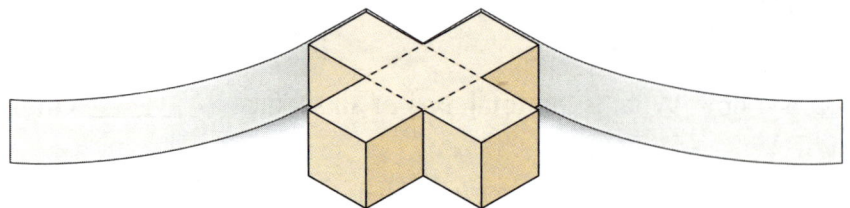

    a How long is the strip of paper? _____cm

    b What is the volume of the shape? _____ cm^3

    c What is the surface area of the shape? _____ cm^2

Extension

PS 1 Here are two different arrangements of eight cubes.

A   B

    a Find the volume and surface area of each.

        A   Volume = _____ cm^3 Surface area = _____ cm^2

        B   Volume = _____ cm^3 Surface area = _____ cm^2

    b Make some other shapes with eight cubes.

        What different surface areas can you find?

    _____

    _____

2 Salim says that if you put several cubes together, the surface area will always be an even number. Try some arrangements of **five** cubes to see if you agree with him. Write down the surface areas you find.

    _____

11.2 Cuboids

In this section you will learn how to:

• find the volume and surface area of a cuboid.

Key words

cuboid
surface area
volume

WORKED EXAMPLE

Calculate the volume and the surface area of this box.

Solution

The box is a cuboid.

Volume = length × width × height
= 6 × 5 × 3 = 90 cm^3.

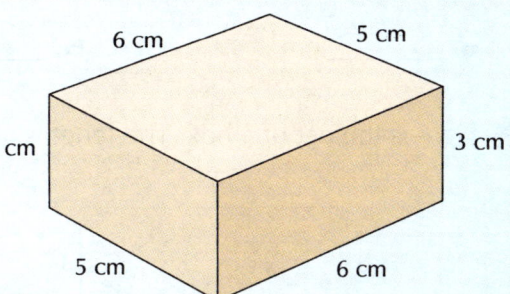

To find the surface area, we need to find the area of each of the six faces and add them together.

The area of the top and the bottom = 6 × 5 = 30 cm^2

The area of the front and the back = 6 × 3 = 18 cm^2

The area of each side = 5 × 3 = 15 cm^2

Total surface area = 2 × 30 + 2 × 18 + 2 × 15 = 126 cm^2 ⟶ Notice that there are three identical pairs of faces.

EXERCISE 11B

1 Find the volumes of these cuboids. They are made from one-centimetre cubes.

a

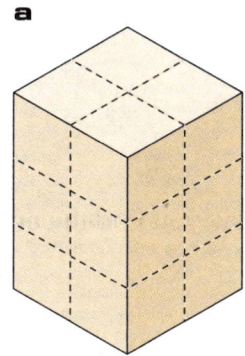

b

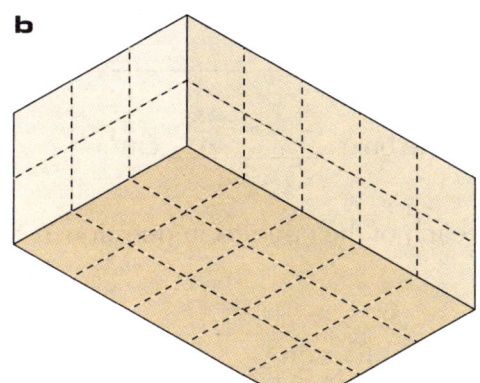

a Volume = _____ cm^3

b Volume = _____ cm^3

2 Find the volumes of these cuboids. All the lengths are in centimetres.

a

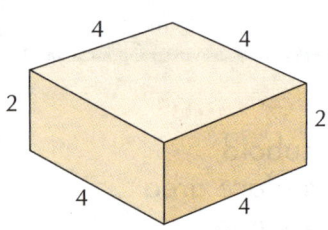

b

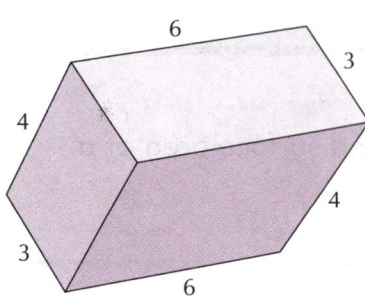

c

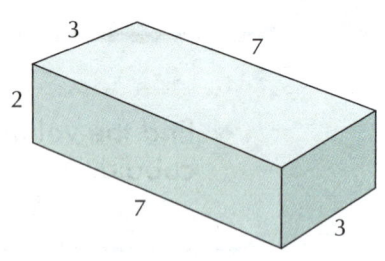

a _____ cm^3 **b** _____ cm^3 **c** _____ cm^3

3 Here is the net of a box. The lengths are in centimetres.

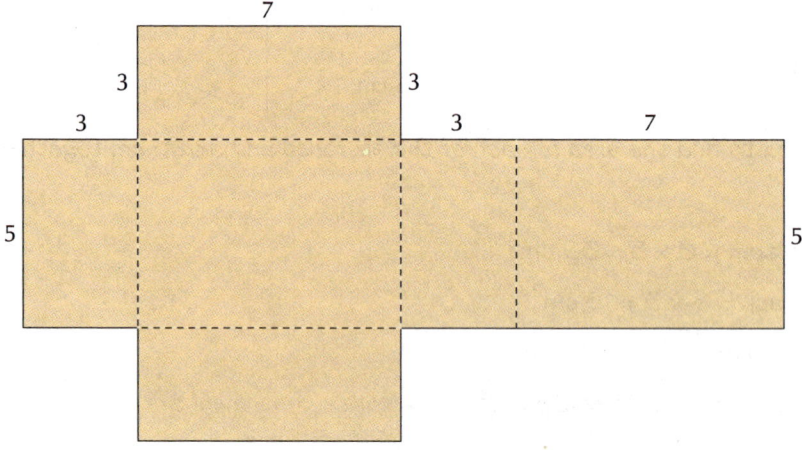

Hint: find the area of the net.

a What is the surface area of the box? _____ cm^2

b What are the length, width and height of the box?

_____

c What is the volume of the box? _____·___ cm^3

4 Find the surface area of each of the cuboids in question 1. Show how you find the answer each time.

a _____ cm^2 Reason _____

b _____ cm^2 Reason _____

5 For large spaces, cubic centimetres are too small and it is more sensible to use cubic metres (m^3).

The floor of a rectangular room is 5 metres long and 4 metres wide. The height of the room is 2.5 metres.

a What is the area of the floor? _____ m^2

b What is the volume of the room? _____ m^3

6 A square piece of paper has sides 10 centimetres long.

A 2-centimetre square is cut out of each corner. The sides are folded up to make a box without a lid.

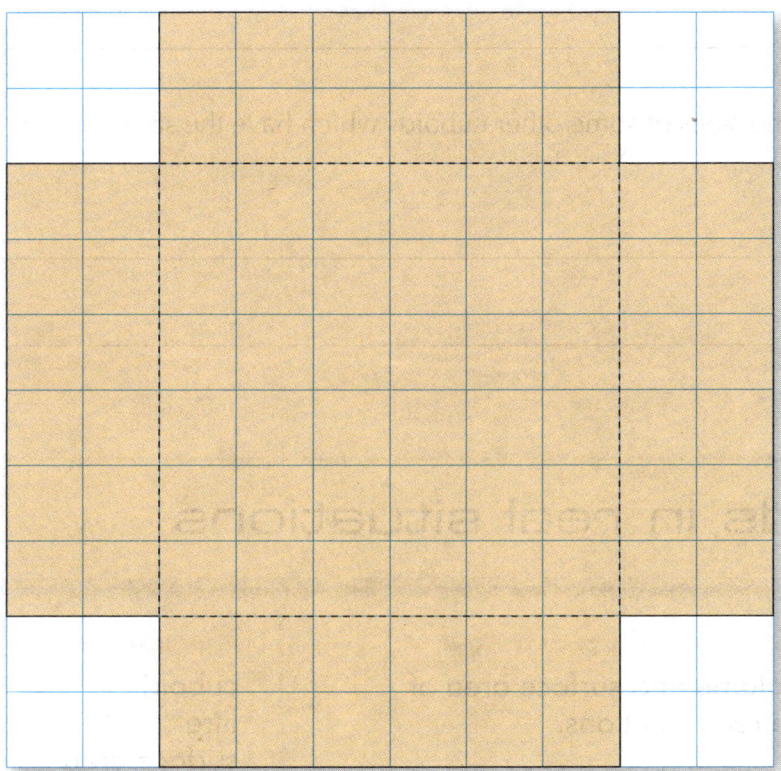

a What is the volume of the box? _____ cm^3

b Investigate what the volume would be if a **different-sized square** were cut off each corner.

_____

_____

_____

7 How many faces does a cuboid have? _____

(Total 1 mark)

AQA, June 2008, Paper 2 Foundation, Question 7a

Extension

AU 1 A cuboid has sides of 2 cm, 2 cm and 15 cm.

a Show that the volume is 60 cm^3.

_____

_____

b Find the lengths of the sides of some other cuboids which have the same volume of 60 cm^3.

_____

_____

11.3 Cuboids in real situations

In this section you will learn how to:
- find the volume and surface area of cuboids in real situations.

Key words
cuboid
litre
surface area
volume

EXERCISE 11C

PS 1 Stock cubes are used in cooking to add flavour to stews. You buy them in a box.

Each stock cube is 3 cm long, 2 cm wide and 1.2 cm high

The box is 10 cm long, 6 cm wide and 1.2 cm high.

How many cubes will fit in the box? _____

M 2 A builder is putting down a concrete floor for an extension to a house.

The floor of the extension measures 4 metres by 3 metres.

A hole has been dug which is 0.5 metres deep.

What volume of concrete must he mix to fill the hole? _____ m^3

M 3 A small swimming pool is 10 metres long, 6 metres wide and 2 metres deep.

a What is the volume of the pool? _____ m^3

b How many **litres** of water are needed to fill the pool? _____ l

Hint: there are 1000 litres in 1 cubic metre.

Extension

U 1 A large matchbox is 12 cm long, 7 cm wide and 2.5 cm high. On the box it says, 'Average contents: 220 matches'.

a What is the volume of the box? _____ cm^3

b Explain why the volume of one match must be less than one cubic centimetre.

_____

The box has a cardboard sleeve with a tray inside.

c What is the area of cardboard in the sleeve? Explain how you calculated this.

_____

d What is the area of cardboard in the tray? Explain how you calculated this.

_____

e Why might your answers to parts c and d not be accurate?

_____

_____

2 Carly's office is 10 metres long, 7 metres wide and 3 metres high.

AU

FM **a** What is the total area of the walls, ignoring any windows or doors? _____ m²

b The walls are to be repainted. One tin of paint is enough for 20 m². The walls will need two coats of paint. How many tins will be required? Explain how you found your answer.

_____

_____

Regulations say that there must be at least 11 m³ of space for every person working in an office.

c What is the volume of Carly's office? _____ m³

d 15 people work in Carly's office. Show whether this breaks the rules or not.

_____

_____

e New carpet is being laid in the office. It costs £16.49 per square metre. What is the total cost of the new carpet?

_____

3 The diagram shows a container in the shape of a cuboid.

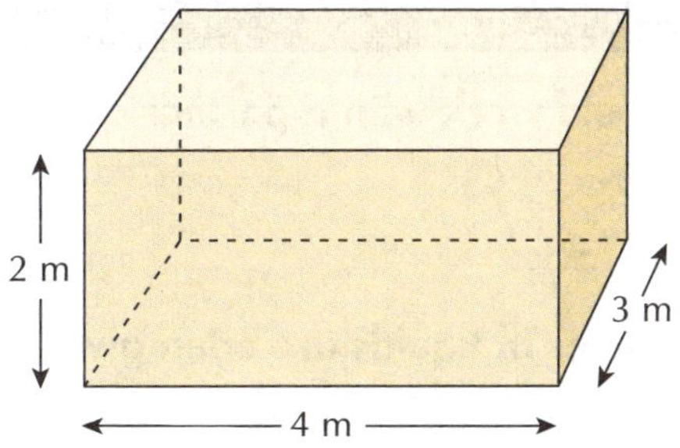

a Work out the volume of the container. State the units of your answer.

Not drawn accurately

_____ (3)

b Ben wants to paint the four outside walls and the top of the container.

One tin of paint covers 6 m^2.

How many tins of paint does Ben need?

You **must** show your working.

Answer: _____ (4)

(Total 7 marks)

AQA, June 2007, Paper 1 Foundation, Question 22

---

checklist

☐ I can find the surface area and volume of shapes made from cubes.

☐ I can find the volume and surface area of a cuboid.

☐ I can find the volume and surface area of cuboids in real situations.

You could work with a partner on this activity.

Useful facts:

Quantities of liquids are often given in litres (l) or millilitres (ml).

1 litre = 1000 ml

1 litre fills a volume of 1000 cm^3.

Imagine a cube where each edge is 10 cm long. That will hold 1 litre of liquid.

1 litre

200 ml

Task A

A supermarket sells its own brand orange juice in two different cartons.

- The **large** carton is 7.5 cm wide, 7 cm deep and 20 cm high. It holds 1 litre.
- The **small** carton is 5 cm wide, 4 cm deep and 11 cm high. It holds 200 ml.

1 Show that the large carton can hold 1 litre of juice.

2 Show that the small carton can hold 200 ml.

The supermarket is interested in the amount of packaging used for the cartons. The manager has asked for your help.

The manager thinks that because the large carton can hold five times as much juice as the small carton, it must need five times as much cardboard to make.

3 What evidence can you give the manager to support his theory or to show that it is incorrect?

Task B

The supermarket plans to start selling the small cartons in a six-pack.

In this pack, six small cartons are held together by transparent plastic film in the shape of a cuboid. This can be done in different ways.

1 The manager has asked you to investigate the different possible arrangements of small cartons in a six-pack.

   How many can you find? Make a sketch of each one to show the manager.

The manager thinks that all the arrangements will use the same amount of transparent plastic film to hold them together because they each contain six cartons.

2 What evidence can you give the manager to support his theory or to show that he is incorrect?

Task C

The manager has asked you to recommend the best way to arrange the cartons in the six-pack. You should take these factors into account.

- The space they will take up on the shelves
- The ease with which they will stack on top of one another without falling over
- The amount of transparent plastic film needed to hold them together
- The visual impact they will have on a customer
- The ease with which they can be carried in a shopping bag

What is your recommendation and what are the reasons for your choice?

Glossary

24-hour clock When time is told in the numbers of hours in the 24-hour day. The morning or AM hours are from 01h00 (the hour after midnight) to 12h00 (noon or midday). Then the afternoon and night, or PM hours start as 13h00, 14h00 … up to 24h00 (midnight).

acute angle An angle that is less than 90°.

add (*add* is a verb; *addition* is the noun) A basic operation of arithmetic, shown by a plus (+) sign. To add, or addition, is the process of combining two or more values to find their total value. For example, 2 + 3 = 5.

angle The space, usually measured in degrees (°), between two straight lines that have a common endpoint (intersecting lines). The amount of turn needed to move from one line to the other.

approximately A value that is not exact, but is accurate enough for the situation. For example, one could say 'approximately 1000 people took part in the survey' rather than '989 people took part'.

arc A curve forming part of the circumference of a circle.

area The measurement of the amount of space a shape occupies.

average A single number that stands for a collection of values. For example, the average of 2 and 6 is 4. (Reason: 2 + 6 = 8, and 8 ÷ 2 = 4)

bearing The direction relative to a fixed point.

calculator An electronic device used for working out mathematical equations.

centimetre A metric unit of length. One hundredth of a metre (0.001). 100 cm = 1 m.

centimetre cube A centimetre cube (cm^3) is equal to the volume of a cube with all sides the length of 1 cm.

chord A line joining two points on the circumference of a circle.

circle A curved line surrounding one central point. Every point on the line is equal distance from the centre point.

circumference The outline of a circle. The distance all the way around this outline.

clockwise Turning in the same direction as the movement of the hands of a clock. (Opposite of anticlockwise.)

concentric circles Circles that have the same centre point.

congruent Exactly alike in shape and size.

cube A solid with six identical square faces.

cuboid A box-shaped, solid object with faces that are all rectangles. It has six flat sides and all angles are right angles.

cylinder A solid or hollow prism with circular ends and uniform (unchanging) cross-section. The shape of a can of baked beans or length of drainpipe.

decimal Any number using base 10 for the number system. It usually refers to a number written with one or more decimal places. For example, $0.1 = \frac{1}{10}$, or $0.01 = \frac{1}{100}$.

diameter A straight line across a circle, from circumference to circumference and passing through the centre. It is the longest chord of a circle and two radii long. (See also *radius*.)

direction The way something is facing or pointing. Direction can be described using the compass points (north, south, south-east, etc.) or using bearings (the clockwise angle turned from facing north).

distance The separation (usually along a straight line) of two points. Can be referred to as length.

divide (*divide* is a verb; *division* is the noun) A basic operation of arithmetic, shown by a division (÷) sign. To split into equal groups or parts, or to 'share fairly'. For example, how can 3 friends share 6 slices of pizza? (6 ÷ 3 = 2 slices each.)

enlargement To make an object larger, in proportion to its original size (also known as a transformation).

estimate To guess a value close to the actual value by using some form of calculation. For example, the student estimated, by counting rows, that there were 360 apples in the box (9 by 10 rows in the top layers and about 4 layers deep = 9 × 10 × 4).

exchange rate Also referred to as 'rate of exchange'. Each country has a unit of currency. When one travels to another country one buys money at the current exchange rate, which is the ratio at which the unit of the currency can be exchanged. For example, during early 2010, the exchange rate for 1 GBP (Great British pound) was approximately ZAR12 (South African rand). In other words, for every pound, one could get 12 rand.

formula (plural: formulae) An equation (numbers and symbols) that shows you how to work out a measurement from other known measurements. For example, the formula for finding the area of a rectangle is Length × Width (or *lw*).

fraction A fraction means 'part of something'. For example, $\frac{1}{5}$. The whole amount is divided into equal parts (in this case 5). A fraction has a number on top (numerator) and a number at the bottom (denominator).

gram (abbreviation: g) A metric unit of mass. (See also *kilogram*.)

kilogram (abbreviation: kg) A metric unit of mass. A bag of sugar may have a mass of 1 kg.

kilometre (abbreviation: km) A metric unit of distance. 1 kilometre = 1000 metres.

kilometres per hour (abbreviation: km/h) The number of kilometres travelled in one hour. For example, the man drove the car at 70 km/h.

line of symmetry A mirror line, or line of reflection; a line that divides something in half so that both sides are the same.

litre (abbreviation: *l*) A metric measure of volume or capacity. 1 litre = 1000 millilitres = 1000 cubic centimetres.

metre (abbreviation: m) A metric unit of length. 1 metre is approximately the arm span of a man. 1 metre = 100 centimetres.

mile An imperial unit of length. One mile is almost 2 km. 1 mile = 1760 yards.

miles per hour (abbreviation: mph) The number of miles travelled in one hour. For example, the woman's cycling speed was 50 mph.

millilitre (abbreviation: ml) A metric unit of volume or capacity. One thousandth of a litre. 1000 ml = 1 litre.

millimetre (abbreviation: mm) A metric unit of length. One thousandth of a metre. 1000 mm = 1 metre.

mixed number A number which is written as a whole number and a fraction. For example, 1 for 1.75.

multiply (*multiply* is a verb; *multiplication* is the noun) A basic operation of arithmetic, shown by a multiplication (×) sign. Multiplication is associated with repeated addition. For example, $3 \times 9 = 9 + 9 + 9 = 27$.

net The shape obtained when a 3-D shape is folded out flat.

obtuse angle An angle that is greater than 90° but less than 180°.

order Arranged according to a rule. For example, ascending order means from lowest to highest.

order of rotational symmetry A shape or pattern can be rotated about a fixed point. If it must be turned through a full circle before the picture looks the same as when it started, it has an order of rotation of one. If it looks the same two or three times during the complete rotation, it has an order of rotation of two or three.

percentage A number written as a fraction with 100 parts. Instead of writing $\overline{100}$ we use the symbol %. So $\frac{50}{100}$ is written as 50%.

perimeter The distance around the outside edge of a shape.

perpendicular At right angles (90°). A line or a flat surface can also be perpendicular to another flat surface.

pint An imperial unit of volume or capacity. 1 pint is just over half a litre. Milk is usually sold in 1-pint or 2-pint cartons. 8 pints = 1 gallon.

pound An imperial unit of mass. 1 pound is 454 g. (Also: 1 pound = 16 ounces, 14 pounds = 1 stone.)

prism A 3-D shape with two identical ends and flat (or perpendicular) sides, all the same length.

proportion Often used in maps drawn to scale. For example, 1 : 300 000. Also, when the ratio of the first two numbers is equal to the ratio of the next two numbers, we can say it is proportional. For example, $4 : 5 = 8 : 10$ or $= \frac{8}{10}$. (See also *ratio*.)

pyramid A 3-D shape made of 4 equal-sized triangles and a square base.

quadrilateral A flat shape, or polygon, with four flat sides. For example: square, rectangle, parallelogram, kite, trapezium.

radius (plural: radii) The distance from the centre of a circle to its circumference (outline).

rectangle A four-sided shape in which all the interior angles are 90°. The opposite sides are of equal length.

reflection The image formed after being reflected. The process of reflecting an object.

reflex angle An angle that is greater than 180° and less than 360°.

rotational symmetry A shape which can be turned about a point so that it coincides exactly with its original position at least twice in a complete rotation. (See also *order of rotational symmetry*.)

scale A scale on a diagram shows the scale factor (or ratio of the length) used to make the drawing. The axes on a graph or chart will use a scale depending on the space available to display the data. For example, each division on the axis may represent 1, 2, 5, 10 or 100, etc. units.

sector A region of a circle, like a slice of a pie, bounded by an arc and two radii.

speed How fast something moves.

square A shape with four equal sides and all the interior angles equal to 90°.

subtract (*subtract* is a verb; *subtraction* is the noun)
A basic operation of arithmetic, shown by a minus
(−) sign. Subtraction is the difference between two
numbers. For example, 25 − 12 = 13.

surface area The area of the surface of a 3-D shape,
such as a cube. The area of a net will be the same as
the surface area of the shape. (See also *net*.)

symmetry When one shape is exactly the same as
another shape if you turn it over, slip it or slide it (that
is, under a transformation), it is said 'to have
symmetry'. For example, the letter T has one line of
symmetry (a mirror down the middle would produce
an identical reflection), the letter N has rotational
symmetry of order two (a rotation of 180° would
produce an image that looks like an N).

tangent A line that just touches the circumference of
a circle at one point without crossing it. Lines,
curves, flat surfaces and curved surfaces can all touch
in this way to form tangents.

tessellation Identical shapes tessellate. They should
have no gaps and should not overlap. Space that is
filled this way, is called a tessellation.

triangle A three-sided shape. The interior angles add
up to 180°. Triangles may be classified as:
1. scalene: no sides of the triangle are equal in length
(and no angles are equal).
2. equilateral: all the sides of the triangles are equal
in length (and all the angles are equal).
3. isosceles: two of the sides of the triangle are equal
in length (and two angles are equal).
4. A right-angled triangle has an interior angle equal
to 90°.

volume The amount of space enclosed by a 3-D
shape. The amount of substance that fills a container.

whole number Numbers that are not negative, or
decimals or fractions. For example, 0, 1, 2, 3, 4 …
49, 580, etc.

William Collins' dream of knowledge for all began with the publication of his first book in 1819. A self-educated mill worker, he not only enriched millions of lives, but also founded a flourishing publishing house. Today, staying true to this spirit, Collins books are packed with inspiration, innovation and practical expertise. They place you at the centre of a world of possibility and give you exactly what you need to explore it.

Collins. Freedom to teach.

Published by Collins
An imprint of HarperCollinsPublishers
77–85 Fulham Palace Road
Hammersmith
London
W6 8JB

© HarperCollinsPublishers Limited 2010

Browse the complete Collins catalogue at
www.collinseducation.com

10 9 8 7 6 5 4 3 2 1
ISBN-13 978-0-00-734009-5

Chris Pearce asserts his moral rights to be identified as the author of this work

British Library Cataloguing in Publication Data
A Catalogue record for this publication is available from the British Library

Commissioned by Priya Govindan
Project managed by Aimée Walker
Edited and proofread by Brian Asbury and Marian Bond
Answer check by Marian Bond
Cover design by Angela English
Concept design by Nigel Jordan
Illustrations by Kathy Baxendale and Gray Publishing
Design and typesetting by Linda Miles,
Lodestone Publishing
Functional maths and problem-solving pages designed and illustrated by Jerry Fowler and edited and proofread by Rachel Faulkner
Glossary by Gudrun Kaiser
Production by Leonie Kellman
Printed and bound by L.E.G.O. S.p.A. Italy

Acknowledgements
The publishers have sought permission from AQA to reproduce questions from past GCSE Mathematics papers.

The publishers wish to thank the following for permission to reproduce photographs. Every effort has been made to trace copyright holders and to obtain their permission for the use of copyright material. The publishers will gladly receive any information enabling them to rectify any error or omission at the first opportunity.

p.14 sonicken/iStockphoto; pp. 16–17 © kavu/iStockphoto, © Viktor Zhugin/Dreamstime.com; pp. 28–29 © Lance Bellers/Dreamstime.com, © Sburel/Dreamstime.com, © Shaw/Wikimedia Commons; pp. 46–47 © Nicky Linzey/ Dreamstime.com, © Jyothi/Dreamstime.com; pp. 54–55 © Monika Adamczyk/Dreamstime.com, © Geraldine Rychter/ Dreamstime.com, © Margaryta Vakhterova/Dreamstime.com, © Elena Schweitzer/Dreamstime.com, © Baloncici/ Dreamstime.com, © Ildar Akhmerov/Dreamstime.com, © Olga Nayashkova/Dreamstime.com, © Darren Fisher/ Dreamstime.com, © Lana Langlois/Dreamstime.com, © Lepas/Dreamstime.com, © Tund/Dreamstime.com, © Roman Ivaschenko/Dreamstime.com, © Monika3stepsahead/Dreamstime.com, © YinYang/ iStockphoto.com, © Virginia Hamrick/iStockphoto, © Robynmac/Dreamstime.com, © Sabina-s/Dreamstime.com, © Yasonya/Dreamstime.com, © Yasonya/Dreamstime.com, © Rcmathiraj/Dreamstime.com, © Linda & Colin McKie/ iStockphoto, © Kieran Wills/iStockphoto, © ALEAIMAGE/ iStockphoto, © Julien Grondin/iStockphoto, © Norman Chan/ iStockphoto, © Jerry Fowler; pp. 68–69 © Jerry Fowler, © Stoke City FC, © Lario Tus/Dreamstime.com, © Richard Thomas/Dreamstime.com, Paul Ellis/gettyimages.co.uk; p.77 © Crown/www.dft.gov.uk, © Honda UK/hondauk-media.co.uk, © Renault/renault.co.uk; pp. 80–81 © Woldee/ Dreamstime.com, © Chris Pearce; p.90 © Crown/ www.hse.gov.uk; p.91 © Oliver Kessler/ iStockphoto.com, © Andrew Zarivny/iStockphoto.com; pp. 94–95 © Niek/ Dreamstime.com, © Astra490/Dreamstime.com, © Clearviewstock/Dreamstime.com, © Jerry Fowler; pp. 106–107 © Daniel Gale/Dreamstime.com, © Jerry Fowler, © Collins Bartholomew Ltd 2009; pp. 118–119 © Jerry Fowler; pp. 130–131 © nikitje/iStockphoto, © Jerry Fowler; pp. 142–143 © Kristian Sekulic/Dreamstime.com, © South12th/Dreamstime.com, © Mackon/Dreamstime.com, © Danny Smythe/Dreamstime.com, © Jerry Fowler; p. 11 answers © R. A. Nonenmacher/Wikimedia Commons.

With thanks to Jan Parry (Secondary Mathematics Consultant, Leicester), Peter Thompson and Annie Sutton (The Angmering School), and Steve Nutting (Oasis Academy, Shirley Park).

Notes

Answers

Chapter 1 Using a calculator

Exercise 1A

1 84

2 816

3 15

4 £217

5 17 weeks

6 The answer is always 1.

7 **a** 1440 **b** 8760

Extension

1 36×54

2 7, 11 and 13

3 Carlos has 263 euros and Juan has 213.

Exercise 1B

1 **a** $\frac{2}{3}$ **b** $\frac{3}{4}$ **c** $\frac{3}{5}$ **d** $\frac{7}{8}$

2 $\frac{3}{6} = \frac{5}{10}$; $\frac{4}{6} = \frac{12}{18}$; $\frac{9}{27} = \frac{7}{21}$

3 Possible answers are $\frac{7}{10}, \frac{14}{20}, \frac{21}{30}, \frac{35}{50}$ and so on.

4 **a** $\frac{5}{6}$ **b** $\frac{5}{8}$ **c** $\frac{7}{12}$ **d** $\frac{7}{8}$

  e $\frac{2}{3}$ **f** $\frac{5}{6}$

5 There are lots of possible answers, including $\frac{1}{3} + \frac{1}{6}$ or $\frac{1}{8} + \frac{3}{8}$.

6 **a** $1\frac{1}{4}$ **b** $2\frac{1}{3}$ **c** $2\frac{3}{4}$ **d** $5\frac{3}{4}$

  e $4\frac{1}{3}$ **f** $2\frac{5}{6}$

7 **a** $1\frac{1}{4}$ **b** $1\frac{1}{6}$ **c** $1\frac{3}{8}$ **d** $1\frac{3}{8}$

  e $1\frac{1}{4}$ **f** $1\frac{3}{4}$

8 **a** The sector from 12 to 3 is $\frac{1}{4}$.

  The sector from 8 to 12 is $\frac{1}{3}$.

  The other sector is $\frac{5}{12}$. Together they make a whole clock face.

9 **a** $\frac{2}{3}$ **b** $\frac{5}{8}$ **c** $\frac{1}{4}$ **d** $\frac{3}{8}$

  e $1\frac{1}{4}$ **f** $1\frac{1}{3}$

10 Top row: $\frac{1}{2}$, $\frac{7}{12}$, $\frac{3}{4}$, 1.

  Second row: $\frac{7}{12}$, $\frac{2}{3}$, $\frac{5}{6}$, $1\frac{1}{12}$.

  Third row: $\frac{3}{4}$, $\frac{5}{6}$, 1, $1\frac{1}{4}$.

  Fourth row: 1, $1\frac{1}{12}$, $1\frac{1}{4}$, $1\frac{1}{2}$.

Extension

1 Because the answer is positive.

2 $\frac{2}{3}$ is bigger because $\frac{2}{3} - \frac{5}{8}$ is positive.

3 **a** $4\frac{1}{4}$ **b** $1\frac{3}{4}$ **c** $2\frac{3}{4}$

Exercise 1C

1 **a** 58 465 **b** 3205

2 oranges = £1.04
grapes = £1.60
total = £2.64

3 Yes. The average is 65.5.

4 **a** £572 **b** £1368 **c** £162

5 £796

6 Option B is £465 cheaper.

7 **a** £0.83 **b** £3.52 **c** £1.92

8   £53.04

Extension

1   Sale prices are £38 for headphones, £118 for the game console and £312 for the television.

2   £342

Problem Solving

Task A

1   28

2   28

3   39

4   46

Task B

12 red beads and 12 blue beads.

Task C

Student's own design and list.

Task D

Student's own costing.

Chapter 2 Units

Exercise 2A

1   **a** cm   **b** ml or l   **c** kg
        d km   **e** g   **f** m

2   **a** 9   **b** 400   **c** 2000
        d 50   **e** 500   **f** 53

3   **a** 2000   **b** 500   **c** 250

4   **a** cm   **b** l   **c** m^2 **d** kg

5   **a** 3500   **b** 4650   **c** 350

6   **a** 1500   **b** 2850   **c** 610

7   **a** 16   **b** 32   **c** 4.8

8   **a** $3\frac{1}{2}$ **b** 14   **c** Less

9   **a** 12.6   **b** 31.5   **c** 63

Extension

1   **a** 1152   **b** 1.152   **c** 4.608

2   Student's own answer. The average weight of a 15 to 16 year-old is just under 9 stones so about 17 teenagers weigh 1 tonne.

Exercise 2B

1   **a** 0815   **b** 1320   **c** 1700

2   **a** 0500   **b** 0230   **c** 0945

3   **a** 1700   **b** 1430   **c** 2145

4   **a** 34 minutes   **b** 9 minutes
        c 20 minutes   **d** 27 minutes
        e 55 minutes   **f** 47 minutes

5   **a** Every half hour **b** 21 minutes
        c 10 minutes   **d** 16 minutes
        e 47 minutes   **f** 24 minutes
        g 41 minutes

6   **a** 20 minutes   **b** 15 minutes
        c 40 minutes   **d** 1 hour
        e 1 hour and 25 minutes
        f 1 hour and 35 minutes

Extension

1   **a** 4 hours and 39 minutes   **b** 1805
        c 0845   **d** 6.05pm, 8.45am

Exercise 2C

1   1.5 km, 5 km and 10 km

2   0.5 litres, 1 pint, 1 litre, 2 pints

3   **a** 88.2 kg and 63 kg
        b 25 kg or 25.2 kg

4   5 and 4

5   92 grams

6   11 gallons

7   **a** 3 hours and 59 minutes
        b 49 minutes, 42 minutes, 1 hour 9 minutes and 1 hour 19 minutes
        c Carlisle and Glasgow

Extension

1   42 km or 41.6 km

2
- a   5.4 m   b   2.85 m   c   1.8 m
- d   0.93 m
- e   Student's own answer. One reason is so that they do not need to use decimal points, which could easily be missed.

3
- a   32   b   0820
- c   It did not stop at Meadowhall

Problem Solving

Task A

1   There are 36 inches in 1 yard.

2   There are 5280 feet in 1 mile.

3   10 chains make 1 furlong.

4   8 furlongs make 1 mile.

5   A furlong is 1/8 of a mile.

6   80 chains make 1 mile.

Task B

1   A furlong is approximately 200 metres.

2   5 furlongs make 1 kilometre.

3   Student's own choice.

4   Student's own choice.

5   1 rod is 5 ½ feet.

6   4 rods make 1 chain.

Task C

Student's own design.

Chapter 3 Scale and drawing

Exercise 3A

1
- a   36 °C   b   39.5 °C   c   40.3 °C

2   Student's own drawing

3
- a   25 mph   b   48 mph   c   63 mph

4
- a   76 cm   b   340 kg   c   87 mph

5
- a   500 ml   b   150 ml   c   650 ml

6
- a   i   97   ii   23
- b   arrow drawn at 560 grams

7   bicycle: 2 metres, bus: 10 metres

8   Probably between 2.5 and 3 metres

Extension

1   350 g; 650 – 350 = 300g; 820 – 650 = 170 g

2   Allow 10 ml either way on these answers: 140, 290, 430, 570, 710

Exercise 3B

1   Cube

2   Cuboid

3
- a   Cube   b   Prism   c   Sphere
- d   Cone   e   Cylinder   f   Pyramid
- g   Cuboid

4
- a   Cylinder
- b   Two circles and a rectangle

5
- a   Cuboid
- b   Student's own drawing.

6   Missing face drawn correctly.

7   A, C, D, F

Extension

1   It folds flat.

2   6 on the top row; 5 and 4 on the bottom row in that order.

3   The net is a sector of a circle with a smaller circle attached to the curved edge of the sector.

Exercise 3C

1   6 miles

2
- a   10 km   b   8 km   c   13 km

3
- a   Student's own drawing.
- b   About 200 miles

4   1 km, 5 km, plus the student's own numbers

5   **a** 16 cm   **b** 13 cm   **c** 8.6 cm
     d 14.4 cm

6   **a** 40, 80, 120, 160
     b **i** 60 km   **ii** 90 km   **iii** 164 km
      iv 116 km

Extension

1   **a** Student's own drawing   **b** 15 km

2   7 cm; half of 14 cm

Exercise 3D

1   **a** 4   **b** 12   **c** 2.5 km
     d 8.5 km   **e** The distance between the lines is 1 km.

2   Long horizontal lengths are 32 cm, long vertical lengths are 17 cm, short horizontal and vertical lengths are 12.5 cm.

3   Student's own drawing.

4   **a**

130	110	90	80	110	100	50
about 80	about 65	about 55	about 50	about 65	about 60	about 30

     b One difference is that in France the speed limit is different on wet and dry roads.

Extension

1   Student's own drawing.

Functional Maths

1   **a** 70m^2

     b Rectangle drawn: 14cm wide, 20cm long
     c Student's own design.

2   Student's own comparison.

Chapter 4 Speed and proportion

Exercise 4A

1   **a** 12 miles   **b** 24 miles
     c 60 miles

2   **a** 50 miles   **b** 100 miles
     c 150 miles

3   **a** 90 miles   **b** 180 miles
     c 360 miles

4

Time (hours)	0.5	1	2	3	4
Distance (miles)	15	30	60	90	120

5   **a** 34 cm   **b** 68 cm   **c** 170 cm

6   30 mph

7   **a** 7 mph   **b** 15 mph   **c** 12 mph

8   **a** 12   **b** 5

9   **a** 70 miles   **b** 210 miles
     c 35 miles   **d** 105 miles

10   8 hours

11   60 km/h

12   **a** 24   **b** 48   **c** 12   **d** 6

Extension

1   **a** going 15 mph, coming back 10 mph
     b 12 mph

2   18 km/h

Exercise 4B

1   **a** butter 180, chocolate 240, flour 60, cocoa 40, sugar 300, eggs 4
     b 12   **c** 270 g

2   **a** 14.50   **b** 290   **c** £344.83

3   **a** 16   **b** 32   **c** 15
     d 80, 60, 160, 320, 250
     e Yes, 1 mile = 1.6 km   **f** 40 000 km

Extension

1   **a** 94p   **b** 85p   **c** Large

2   Small pack. Each glass costs 80p whereas each glass in the large pack costs 85p.

3   **a** 60%   **b** 210 ml

Functional Maths
Starter:
325 g potatoes
20 g plain flour
65 g peas
65 g carrots
1 small onion
half–1 red chilli (depending on how hot you like it)
1 clove of garlic
$\frac{1}{2}$ teaspoon of cinnamon
$\frac{1}{2}$ teaspoon of cumin
Juice of $\frac{1}{2}$ of a small lime
1 tablespoon of chopped coriander
Salt and oil

Main
600 g aubergines
1 large onion
300 g minced lamb
2 medium eggs
Salt and pepper

Dessert
80 g plain flour
40 g wholemeal flour
1 large egg or 2 medium eggs
$\frac{1}{4}$ teaspoon of vanilla essence
$\frac{1}{2}$ teaspoon of ground cinnamon
3 small Cox's apples
4 tablespoons dry cider

Chapter 5 Area and perimeter

Exercise 5A

1   **a** 18   **b** 28

2   **a** 120m   **b** 170 m   **c** 150 m

3   **a** 20 cm   **b** Student's own drawing.

4   **a** 16 cm   **b** 18 m   **c** 36 mm

5   **a** 6 + 12 + 5 + 9 + 7 = 39
    b 7 cm   **c** 6.5 cm

6   20 m

Extension

1   Triangle 12 cm, square 9 cm

2   160 cm

Exercise 5B

1   **a** 20   **b** 22

2   31 km^2 to 37 km^2

3   **a** Yes. The perimeter of the triangle is six small triangle sides so is the perimeter of the hexagon.
    b No. The large triangle is four small triangles, the hexagon is six.

4   **a** 15 cm^2 **b** 20 m^2 **c** 17 mm^2

5   **a** 24 cm
    b 32 cm^2

6   **a** 72 m^2 and 20 m^2 **b** 92 m^2

Extension

1   124 m^2

2   **a** 40 cm^2 **b** 20 cm^2

3   **a** 9 cm^2 **b** 28 cm

Exercise 5C

1   **a** 210 feet   **b** 2106 square feet.

2   **a** 228 feet   **b** 2808 square feet.

3   No. 2808 is not double 2106.

4   **a** Singles perimeter is 122 feet and area is 748 square feet.
    b Doubles perimeter is 128 feet and area is 880 square feet.

5   **a** 320 yards
    b 6000 square yards

6 **a** The area
    b Perimeter = 17.2 m, area = 17.85 m^2
    c To fit an edging border is one reason.

7 **a** Area = 24 m^2; 24 × £8.49 = £203.76
    b Perimeter is 20 m, so 8 pieces are needed. £20 × 8 = £160
    c No, you still need 8 pieces because doors are generally less than 1 metre wide and the skirting board is sold in lengths of 2.5 metres.

Extension

1 Along the 3.3 m side you need 7 tiles.
Along the 4.8 m side you need 10 tiles.
7 × 10 = 70
70 × £3.49 = £244.30

Functional Maths

Task A

1 The maximum area of a senior pitch is 10 800 m^2.
The maximum perimeter of a senior pitch is 420 m.

2 **a** The minimum size for a senior pitch is 90 m long and 45.5 m wide.
    b The minimum area of a senior pitch is 4095 m^2.
The minimum perimeter of a senior pitch is 271 m.

3 There is a big difference.
The maximum-sized pitch is more than twice the area of the minimum one.

Task B

1 Student's own scale drawings.

2 The table below shows the areas and perimeters of the pitches (rounded to the nearest whole unit):

Pitch	Maximum		Minimum	
	Area (m^2)	Perimeter (m)	Area (m^2)	Perimeter (m)
Seniors	10 800	420	4095	271
Under 16	6438	329	3745	256
Under 14	5096	294	3312	237
Under 12	4163	266	2867	221

3 Student's own comparisons.

4 Student's own recommendation.

Chapter 6 Symmetry

Exercise 6A

1 **a**  **b**

c    **d**

e    **f** no lines of symmetry

2 **a** 4   **b** 2   **c** 8

3 **a**    **b**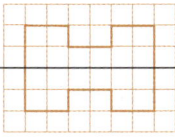

4 **a** 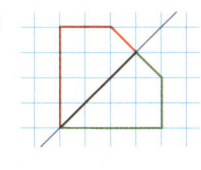   **b**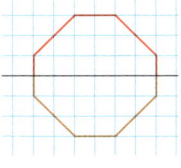

      pentagon   octagon

5 **a** S and P **b** A and E **c** H

6 The missing numbers are 702; 728 and 756; 754, 783 and 812

Extension

1 **a** and **b** 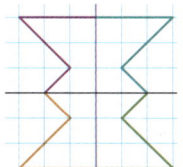   **c** 2

2 **a**

    b Any word using B, C, D, E, H, I, O or X

Exercise 6B

1 **a** 2   **b** 2   **c** 4   **d** 5
    e 1   **f** 3   **g** 1   **h** 2

2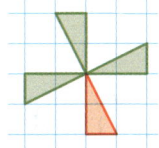

3 **a** F   **b** X

4 **a** lines drawn correctly
    b square added to bottom right-hand side

5

	Order of rotational symmetry	Lines of symmetry
S	2	0
N	2	0
W	1	1
X	2	2
Z	2	0

Extension

1 A number of possible answers.

2 Student's own drawing.

Exercise 6C

1 **a** Square **b** Parallelogram
    c Square and parallelogram
    d and **e** Student's own shapes.

2 **a** 1 line, no rotational symmetry
    b 1 line, no rotational symmetry
    c 3 lines, rotational symmetry of order 3
    d No lines of symmetry, rotational symmetry of order 3
    e 1 line, no rotational symmetry
    f 2 lines, rotational symmetry of order 2

3 **a i** No lines, no rotational symmetry (rotational symmetry of order 2, if you ignore the shading)
    ii 1 line, no rotational symmetry
    b Student's own drawings and descriptions

Extension

1 Student's own pattern.

Functional Maths

Task A

Hubcap	Order of rotational symmetry	Lines of Symmetry
1 Vauxhall	9	9
2 Ford	8	8
3 Hyundai	5	0
4 Ford	2	2
5 Vauxhall	7	7
6 Renault	10	0
7 Citroen	14	14
8 Almeira	6	0
9 Saab	3	3
10 Ford	8	0
11 Audi	5	5
12 Renault	9	0
13 Smart	6	6
14 Ford	8	8
15 Fiat	5	0

Task B

Student's own design.

Chapter 7 Angles

Exercise 7A

1 A and C

2 **a** obtuse **b** acute **c** reflex
       d right

3 The top two

4 B, E, D, A, C

5 **a** 20° **b** 35° **c** 80°
       d 120° **e** 165° **f** 270°

6 Student's own drawing

7 Student's own drawing

Extension

1 **a** Student's own drawing **b** no

2 Student's own drawing

Exercise 7B

1 $a = 90°$, $b = 135°$, $c = 56°$

2 $a = 120°$, $b = 205°$, $c = 120°$

3 **a** **i** Angle 78° is acute, 144° is obtuse
          ii 138°
       b No. They add up to 190° not 180°.

4 $a = 30°$, $b = 35°$, $c = 110°$,
       $d = 20°$

Extension

1 **a** 75°, 130°, 100°, 55°; the total should be close to 360°
       b It is true
       c Student's own drawing

2 The spiral, star and flower are 3 angles on a straight line, so must add up to 180°. Each triangle has angles of a spiral, star and flower. So the angles of each triangle must add up to 180°.

Exercise 7C

1 **a** Student's own drawing **b** 15°
       c Student's own check

2 **a** 3 equilateral triangles.
       b 30°, 60° and 90°
       c Triangles are rigid.

3 3 angles make 360°. Each one is 360 ÷ 3 = 120°

Extension

1 135° = 90° + 45°

2 **a** 2 × 120° + 2 × 60° = 360°
       b Student's own drawing.

Problem Solving

Task A

2 Student should have drawn an equilateral triangle with a line down the middle. All angles are 60°.

3 Student should have drawn a triangle with two 30° angles and one 120°

angle. There should be a line bisecting the 120°.

4 There are three ways to do this (including a rectangle).

Task B

1 There are two shapes. One has angles of 30, 90, 90 and 150 degrees. The other has angles of 60, 90, 90 and 120 degrees

2 Student's own answer.

3 Student's own answer.

4 The total for a four sided shape is 360 degrees.

Chapter 8 Circles

Exercise 8A

1 **a** 3.5, 2.5, 2 cm
  b Student's own drawing
  c 7, 5, 4, 6 cm

2 **a** Diameter drawn **b** Tangent drawn
  c **i** Midpoint marked and labelled
  ii It is a right-angle

3 **a** Diameter **b** Radius
  c Tangent **d** Radius
  e Chord **f** Centre

4 **a** 4 cm **b** 110°
  c line of symmetry drawn
  d tangent at *A* drawn
  e chord at *AB* drawn

5 **a** 6 cm **b** 3 cm

6 Student's own drawing.

Extension

1 Student's own drawing.

Exercise 8B

Answers in this exercise are approximate. Values close to these are also correct.

1 **a** 31 cm **b** 16 mm **c** 25 m
  d 31 km **e** 19 mm **f** 22 m

2 **a** 19 cm **b** 28 m **c** 12.5 cm

3 **a** Between 6 and 7 cm
  b Student's own drawing.
  c Hard to measure if you cannot get at the end of the pipe.

4 Between 4 and 5 metres

Extension

1 **a** $31\frac{3}{7}$ cm or between 31 and 32 cm
  b It should be the same.

2 **a** 62.8 metres
  b Missing numbers are 15.7, 62.8 and 125.6

3 About 31 cm

4 **a** About 12 500 miles
  b About 8000 miles

Exercise 8C

1 **a** 1p 20 mm; 2p 25 mm; 5p 18 mm; 10p 24 mm; £1 22 mm; £2 28 mm
  b No, a 2p is larger than 5p and 50p is larger than £1.

2 Student's own drawing

3 **a** Square packing
  b Student's own drawing
  c Student's own drawing
  d Hexagonal

Extension

1 **a** About 24 cm
  b Width 10 cm. Length at least 24 cm and probably longer to give an overlap.

Problem Solving

Task A

1 Student's own drawing

2 Student's own drawing

3   Student's own drawing.

4   **a**   Every point on the line is equal distance (equidistant) from A and B.
   b   All the places that fall on the same side of the line as A will be nearer to transmitter A than transmitter B.

Task B

1   Student's own drawings.

2   **a**   The points form a circle of radius 40 km. The centre of the circle is 80 km from A and 20 km from B.
   b   Inside the circle.

Chapter 9 Transformations

Exercise 9A

1   C, because it is a different size.

2   **a**   A + E, B + F, C + D
   b   B, C, D and F

3   **a**   A + H + L, B + C, E + F, G + I + J + K
   b   D, because it is not congruent to any other shapes.

4   **a**   Trapezium   **b**
   c   They are the same size and shape. One could fit on top of the other if it were turned over.

Extension

1   **a**   They are different sizes.
   b   Student's own measurements. purple : orange = 1 : 2
   c   Student's own drawing. Perpendicular sides should be 6 and 8 squares long.

2   Student's own drawing.

Exercise 9B

1   Student's own drawings

2   **a**   A and E   **b**   C and D   **c**   8 cm^2

3   **a**   Student's own drawing   **b**   Yes
   c   Perhaps they would be more expensive or difficult to make.

Extension

1   Student's own drawings.

2
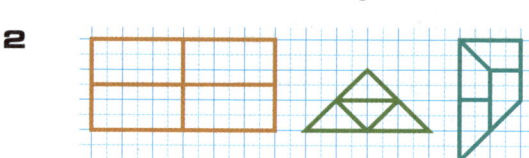

3   Student's own drawing.

Exercise 9C

1
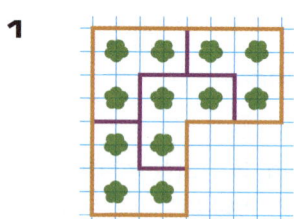

2   Student's own drawing.

3   Student's own drawing.

4   Student's own drawing.

Extension

1   Student's own drawing.

2   **a**   Student's own drawing.
   b   3 lines of symmetry and rotational symmetry of order 3.
   c   Student's own drawing.

Problem Solving

Task A

1   A and I
B and C
D, J, G and K
E and F
H and L are not congruent to any other pentominoes.

Task B

1   Please see diagram below.

2   Student's own answer. 6 of the pentominoes should look the same when turned over.

3   Student's own drawing.

Task C

Student's own answers.

Chapter 10 Constructions

Exercise 10A

1   **a** 6.5 cm   **b** 8.4 cm   **c** 11.8 cm

2   9.4 cm, 6.3 cm, 35°

Extension

1   **a** 8.3 cm   **b** 11.3 cm   **c** 6.2 cm

2   Student's own drawings.
Third angles should be:
a 75°   **b** 67°   **c** 65°   **d** 110°
e 42°   **f** 60°

3   **a** Isosceles   **b** Both 6.7 cm
c Equilateral   **d** 7.5 cm
e Student's own check.

4   Student's own drawings.

5   Student's own check.

6   **a**   Student's own drawing.

      b   It is impossible because 4 + 6 is less than 11

      c   Student's own drawing.

7   Student's own drawing.

Exercise 10B

1   **a**   105°   **b**   150°

2   hill 075°, mosque 155°, wood 217°, church 325°

3   **a**   080°   **b**   260°

Extension

1   **a**   25 miles   **b**   030°

      c   Student's own drawing.

      d   Student's own drawing.

Exercise 10C

1   Clockwise from north, 000°, 045°, 090°, 135°, 180°, 225°, 270°, and 315°

2   Student's own drawing.

3   **a**   075°   **b**   650 km   **c**   255°

4   **a**

	Bearing	Distance
Glasgow to turn	165°	170 miles
turn to Manchester	100°	40 miles

      b   To avoid planes travelling in the opposite direction from Manchester to Glasgow.

Extension

1   Student's own drawing.

Problem Solving

Student's own activity.

Chapter 11 Volume

Exercise 11A

1   **a**   4 cm   **b**   4 cm^3 **c**   18 cm^2

2   **a**   4 cm^3 **b**   6   **c**   16 cm^2

3   Student's own answer

4   **a**   A: 6 cm^3 and 22 cm^2

          B: 6 cm^3 and 24 cm^2

      b   Student's own answer.

5   **a**   12 cm   **b**   5 cm^3 **c**   22 cm^2

Extension

1   **a**   A: 8 cm^3 and 24 cm^2

          B: 8 cm^3 and 34 cm^2

      b   Student's own answer.

2   Student's own answers.

Exercise 11B

1   **a**   12 cm^3 **b**   30 cm^3

2   **a**   32 cm^3 **b**   72 cm^3 **c**   42 cm^3

3   **a**   142 cm^2

      b   length = 7 cm, width = 5 cm, height = 3 cm   **c**   105 cm^3

4   **a**   32 cm^2, 2(2 × 2) = 8, 4(3 × 2) = 24, 8 + 24 = 32

      b   62 cm^2, 2(2 × 3) = 12, 2(2 × 5) = 20, 2(3 × 5) = 30, 12 + 20 + 30 = 62

5   **a**   20 m^2 **b**   50 m^3

6   **a**   72 cm^3 **b**   Student's own answer.

7   **a**   6

Extension

1   **a**   2 × 2 × 15 = 60

      b   Student's own answers.

Exercise 11C

1 10

2 6 m^3

3 **a** 120 m^3 **b** 120 000 litres

Extension

1 **a** 210 cm^3
   b Because 220 matches can fit in a box with a volume of 210 cm^3.
   c 228 cm^2, 2(12 × 7) = 168, 2(12 × 2.5 = 60, 168 + 60 = 228
   d 179 cm^2, 12 × 7 = 84, 2(2.5 × 7) = 35, 2(2.5 × 12) = 60, 84 + 35 + 60 = 179
   e The card needs to overlap and to be joined. The tray needs to be slightly smaller than the sleeve.

2 **a** 102 m^2
   b 11 plus student's own explanations
   c 210 m^3
   d No it does not; 15 × 11 = 165
   e £1154.30

3 **a** 24m^3 **b** 7

Functional Maths

Task A

1 The volume of the large carton = 7 × 7.5 × 20 = 1050 cm^3. 1 litre = 1000 cm^3. Therefore the carton can hold 1 litre.

2 The volume of the small carton = 5 × 4 × 11 = 220 cm^3. This is more than 200 ml so the carton can hold 200 ml.

3 Make a net of the carton to work out the area of the cardboard needed to make a carton. The area of cardboard for the large carton = 685 cm^2. The area of cardboard to make a small carton = 238 cm^2.

Area for large carton ÷ area for small carton = 685 ÷ 238 = 2.88
The large carton needs less than 3 times the amount of cardboard the small carton requires. This shows that the manager's theory is incorrect.

Task B

1 Student's own investigation.

2 The surface area is the amount of transparent plastic film it will take to enclose the six-pack.
Calculate the surface area of a few of the arrangements to show the manager that the different arrangements will require different amounts of transparent plastic film. This will show that the manager's theory is incorrect.

Task C

Student's own recommendation. Student should calculate the surface area of the base of each arrangement to find the one that takes up the smallest amount of space on the shelf. Reasons behind the recommendations include these points.

- Stacking is easier with larger base area compared to height.

- The amount of transparent plastic film is determined by the surface area.

- The visual impact is highest when the largest faces of the cartons face the customer when displayed on the shelves.

- The ease of carrying in a shopping bag is minimised by reducing the 'bulkiness' of the six-pack. Minimising all three dimensions (width, height and depth) will therefore be optimum.

Notes